RE-IMAGINING
MATHEMATICS

To the memory of my father,

Arun Sen

RE-IMAGINING MATHEMATICS

Learning through the Magic of Nature, the Arts and Friendship

ASHNA SEN

PENMAN BOOKS
A BOOK PUBLISHING COMPANY

PENMAN BOOKS

Office No. 303, Kumar House Building,
D Block, Central Market, Opp PVR Cinema,
Prashant Vihar, Delhi 110085, India
Website: www.penmanbooks.com
Email: publish@penmanbooks.com

First Published by Penman Books 2021

Title: Re-Imagining Mathematics
Price: ₹599 | $15
ISBN: 978-93-90617-91-3

CONTENTS

WORDS OF PRAISE

In her engaging book 'Re-Imagining Mathematics,' Ashna Sen argues for conceptualising mathematics through the wonders and patterns of nature that any child is naturally drawn to and intuitively feels. Because the trees, rivers, planets and indeed our brains themselves exhibit order and patterns that are already familiar to early learners, mathematics becomes a delightful recognition of what we already know and what we already are, rather than an abstract and separate study. This book is an absolutely delightful re-recognition of the innate connection between numbers, ourselves, and our natural world.

—Dr. Michelle Johannes,
Computational Research Physicist,
Naval Research Laboratory, USA

~ ~ ~

In an often intimate, autobiographical atmosphere, Re-Imagining Mathematics explores the wonders of number in relation to progressive pedagogy, the natural world and physics, sacred geometry and the arts, and even human companionship. This profoundly lucid book seems to me to enshrine a delicious paradox, in itself quite appropriate to mathematics, for it combines concision with infinity. Mathematics is of course central to it, but taken in an ultimate sense: as a key to encountering the living mystery that we inhabit as individual humans, and which indeed constitutes us.

—**Dr. Toby Mayer,** *Senior Research Associate, Institute of Ismaili Studies, London*

~ ~ ~

This book is an enjoyable and rewarding read - that lifts the spirit as well as informs the mind. The notion of teaching mathematics using examples from Nature as part of a holistic education is illustrated with fascinating examples of e.g. the Fibonacci series in flower petals and fractals in tree branches, rounded out with 'mathematical haikus' and apposite references to art and music. Perhaps only schools such as Brockwood where the author has taught would give teachers the freedom to innovate like this, but if some of those who read this book were to become maths teachers, then hopefully this spirit will come through in their teaching. And the author would surely consider that the best possible reward for her labour of love!

—**Profs. Amanda & Subir Sarkar,** *Department of Physics, University of Oxford*

~ ~ ~

Dr. Sen's witty, perceptive, energetic interdisciplinary aptitude and her ability to frame her thoughts in a lucid, relaxed autobiographical style invite us on a re-oriented journey towards the heart of being, where learning maths coincides with imagining, where knowledge is gathered only as an attendant to friendship with and reverence to things and to beings, where playfulness is always serious and seriousness slippery and playful.

—**Dr. Valentin Gerlier,** *Research Associate, St. Edmund's College, Cambridge University, UK*

~ ~ ~

If you're like me and feel that you have little or no relation to numbers and maths, this book does absolute wonders in demonstrating how the quality of mathematics pervades everything in ways that are meaningful to who we are as humans. Like viewing the world through a different lens, I couldn't help but see the magic of numerical relations in nature and design.

—**Dr Todd S. Mei,** *Philosophy2u,*
Former Head of Philosophy,
University of Canterbury, Kent, UK

~ ~ ~

For three hundred years we have lived under the delusion of bringing nature under human mastery, driven by a fear of uncertainty disguised as courage. For many peoples around the world now the time has come to abandon that delusion which has only brought a deep sense of social and cultural alienation. I commend this fine book with enthusiasm. I believe it will strike a deep note of recognition in its readers, and contribute to the new community of learning that is spontaneously emerging around the world.

—**Dr. Joseph Milne,** *Temenos Academy, UK*

~ ~ ~

Dr. Sen's wondrous treasure of a book will open many eyes and hearts to that which has been lost - the sacred majesty of number and how mesmerising and joyful learning can and should be! This re-imagined mathematics returns number to the forest and Art to natural proportions, a fine organic medicine for our fractured world.

—***Kat E. Erikson,*** *Children's Author*

~ ~ ~

We have long known that education isn't restricted to the classroom. This wonderful book bears witness to the materialisation of maths in our everyday lives: the maths of our footprints and forests, our bodies and buildings, our reasons and relations. Through fertile examples of pattern, form, rhythm and colour, maths is shown to constitute our environment, thereby forming who we are. The rich scholarship is seamlessly interwoven with profound and personal mathematical insights, bringing us back to what we once knew well, but, in much contemporary education at least, seem to have forgotten: that when taught well, maths is the most life affirming dimension of a human curriculum.

—**Dr. David Lewin,** *Senior Lecturer in Philosophy of Education, University of Strathclyde, UK*

ABOUT THE AUTHOR

Ashna Sen was born and brought up in Calcutta. She double majored in physics and mathematics from Mount Holyoke College, USA, and thereafter obtained an MSc in geophysics from Stanford University. She has a PhD in mathematics from the University of Manchester, UK, and completed postdoctoral research from the University of Ljubljana in Slovenia. She has also studied and done research at Princeton University, Brookhaven National Laboratory and Oxford University. Ashna taught mathematics to undergraduates at St. Xavier's College, Calcutta, and in a mathematics village in Şirince, Turkey, but became interested in innovative mathematics education during her years at Brockwood Park School, Hampshire, UK. Over the years, she has contributed to several journals and magazines, including the Journal of Krishnamurti Schools and The Temenos Review. A historical memoir written by her on the lives and times of civil servants in Bengal during the pre-partition era of the British Raj, called *The Rusted Trunk*, was published by Writer's Workshop, Calcutta, in 2011, and later translated into Bengali.

For the past decade Ashna has been living in the steep climes of the eastern Himalayas with her husband and two children. *Re-imagining mathematics: learning from the magic of nature, the arts and friendship* (Penman Books, 2021) is her second book.

ACKNOWLEDGMENTS

Foremost, I am grateful to Mita Radhakrishnan and Tapas Desrousseaux for insisting that I write this book and for their faith and belief in my abilities. What started off as friendly scribbles on a blank page has turned into this book length document because of the persistence and encouragement of the following, whom I must thank at the very outset – Jyotsna Ramachandran, Nikhil Raval, Bernard Ginsbourg, Charles Daehler and Toby Mayer.

To Joseph Milne and Valentin Gerlier I owe a special debt of gratitude for their invaluable input in shaping my thoughts, for their enthusiastic endorsement of my ideas, and for happily agreeing to write the foreword and afterword for this book.

Yusuf Simick, Norden Pempahishey and Mingma Tshering Bhutia, came to our rescue when my family and I were caught unawares during the first lockdown of 2020. For their hospitality and support during those eight months when much of this work was completed, I am forever obliged. The teachers of Mayel Lyang Academy, Kalimpong, have been a source of inspiration throughout the writing of this book – and many of the examples and topics covered here were first shared with them; thank you for being my experimental subjects!

Sukanya Ray and Krishno Dey opened their hearts and home to us at a precarious time during the summer months of 2021. For your valuable suggestions and a comprehensive introduction to batik printing, Sukanya, and for egging me on to finish this project, many thanks.

I also wish to acknowledge Harriet Pollatsek, Benedicte Notteghem, Robin Gupta, Krishna McKenzie, Murshed Nawaz, Biswarup Choudhury, Anna Louisa Taylor, Chhaya Yadav, Mark Peterson, Radha and Jayanta Choudhury, Kristen Edwards, Lisa Romney, Leeanne Edwards, Wendy Johnston, Shehla Khan, Imtiaz Ahmad, Shaida Nabi, Myriam Blin, Michelle Johannes, Salma Ali, Kaushik Sarkar and Jayeeta Verma Sarkar, Abhik Samanta, Bhaswati Bhattacharya, Pradip and Nandita Bhattacharya, Abhijeet Sen, Javed Kutty, Sayeed Ahmad Baba and Suraiya Baba for their continual support, friendship and guidance through the years.

To my publishers, Penman Books and their team and Tarun Singh in particular, I say a special thank you for the meticulous care they took in bringing this published version into fruition.

Todd Mei, Kat E. Erikson, David Lewin, Michelle Johannes, Subir Sarkar and Amanda Cooper-Sarkar offered their encouraging feedback and suggestions, which gave me the confidence that I was on the right track. I am grateful to them for their attention and advice.

My relatives and friends across three continents have propped me up during difficult times. I feel deeply privileged to have received their care and reassurance and take this opportunity to express my sincere gratitude to them.

Finally, I wish to thank my family: my husband Brian Edwards and my children for their love, appreciation and help in ways too many to enumerate.

FOREWORD
BY DR. JOSEPH MILNE

Aristotle says the mind is inclined by nature towards truth. It follows from this that we enjoy beholding things. We delight in simply seeing and hearing things, in attending. Yet this simple attending gives rise to wondering about the nature of things, in particular about the underlying order of everything. We have an intuition of 'the whole' within which we experience all the variety of things as part of some great order. The mind is drawn to this great order. And from this arise all the branches of knowledge which seek to understand that order in different ways.

But more than this, the mind desires to affirm the truth of things. And so the arts and religion are spontaneously born as natural aspects of human life and society. This affirmation is found in the earliest historical documents, which are beautiful hymns to the gods. To wonder and to praise truth with beauty is the most natural thing, and where it prevails civilisation flowers. Justice and goodness are honoured, and so human life in community flourishes.

So, from the natural inclination of the mind towards truth, human life flourishes. That is the natural process. There follows no end to the insights into the mystery and order of things. To enquire, simply out of the desire to know and to acknowledge, leads intelligence to honour truth wherever it is found. This honouring truth, which comes from wonder, puts us in the right relation to things, and to the world at large. It leads naturally to living virtuously. In a mysterious way the order of things leads to a perception of justice. This is easily observed in young children. They desire to play in justice. Without justice a game is not a game. In the same way, learning and enquiring is not true learning and enquiring if it is done without honouring.

The ancient philosophers observed this remarkable connection between exploring the order of things and justice, in honouring and living justly. Study and ethics belong together. And so the natural quest for knowledge leads to virtue. One naturally wishes to live in harmony with the nature of things. From this desire arises discourse on the nature of society, within the context of the great order of things, the cosmos.

The capacity to reflect on truth and justice distinguishes the human species from all the other species. It is a great gift that nature has conferred upon us. It means that not only can we learn directly from nature but we can care for it and guard it from harm. Through the crafts and skills we have learned from nature we have more than sufficient to sustain life. We are thus free to pursue greater things and to put beauty into everything undertaken.

This rather wonderful book which Ashna Sen has written shows how knowledge may be approached in this way. Mathematics is one of the foundational ways in which nature speaks to our intelligence and discloses its astounding order. We see that mathematics is a real subject, one that even young children will see in a leaf or a flower. In a remarkable way, nature shows herself in number and pattern and proportion in every living thing, in the silence of the snowflake or the shining crystal. The light of the sun in a dewdrop shows they were made for each other. The poet and the mathematician wonder at the same mystery and share in the same affirmation.

Not only do we find nature 'singing numbers', so to speak, but number gives rise to the many arts and in particular to architecture – the master art because it houses all the other arts. People flock to see the beautiful cathedrals and temples that were built with great knowledge of geometry. We intuitively recognise the marvellous beauty of proportion and the way space is made a dwelling place for contemplation.

But also, numbers in themselves are marvels, each number having a character of its own and a way of working with the other numbers. This becomes audible in the musical harmonies and dissonances. For Pythagoras and Plato, the numbers are worth contemplating in themselves. And from this there have arisen mystical traditions of numbers. Here we see that mathematics is not right if merely used to measure and explain things, as mere calculation. Numbers are more properly studied in themselves, and their practical application is a secondary gift they confer.

There is another side to this. Where the truth of things is pursued with honour and reverence, nature discloses aspects it otherwise keeps hidden. This is not so strange. Only the musician who truly loves music for itself becomes a great musician. Love of a subject brings the intelligence into a right relationship with it. Nothing is more natural than that, and certainly all children should be encouraged to study what they love. It will take them on their right path in life and enable them to contribute to the common good.

For all kinds of misguided reasons learning in our times is no longer seen as natural. And so it has been removed from nature and the wonders nature shows us by itself. It has been regimented into artificial structures and ugly classrooms where nothing invites thought or enthusiasm. And instead of being undertaken for the sake of knowledge in itself, it is reduced to mechanical acquisition of facts which are held to be needful in the life of commerce. And the schools have to compete with each other to climb the league tables. So education is itself entirely secondary to the aims of schooling. Such an instrumental approach to education breaks the link between truth and ethics and dishonours human nature. The child is practically reduced to a commodity on a production line. Most teachers know this is wrong, while some 'smuggle in' a bit of real education where they can.

To break away from such a misguided system, informed by such misguided theories, requires considerable courage. But courage here is built on insight, insight into what is natural and what moves with the course and telos of nature. Children naturally take to learning amongst nature, as we see in the forest schools described in Chapter 2. In the forest schools children become social as a matter of course. Learning with their peers is an adventure.

The wonderful insights of this book call out for realigning education with our natural human orientation towards the truth of things. Our age has become so accustomed to seeing the world through the prism of technology and reductive materialist theories that nature, including human nature, has almost become invisible to us. We consult concepts instead of actualities, ideologies instead of virtue, and market exchange values instead of the direct gifts of providence.

Behind these modern strategies lies a forgetfulness and distrust in our natural human way of being. The ancient philosophers understood that there is a direct correspondence between our human reason and the order of the cosmos – that Nature is herself intelligent and just. For three hundred years we have lived under the delusion of bringing nature under human mastery, driven by a fear of uncertainty disguised as courage. For many peoples around the world now the time has come to abandon that delusion which has only brought a deep sense of social and cultural alienation. There is a desire for restoration, for living more authentically, and for society founded in friendship. The most fundamental approach to such a restoration is through rediscovering the roots of knowledge in the relation of our human intelligence with the intelligence of nature.

I commend this fine book with enthusiasm. I believe it will strike a deep note of recognition in its readers, and contribute to the new community of learning that is spontaneously emerging around the world.

—***Joseph Milne,*** *Canterbury UK*

CHAPTER ONE

THE LANGUAGE OF MATHEMATICS

"The power of mathematics is often to change one thing into another, to change geometry into language."

—Marcus du Sautoy

Ever since I was a child I have wondered why is it that language comes to us all naturally. Especially when we are young, without much effort, we pick up our mother tongue as if it were already 'there' in some form, within us. Why do we even use language as a human race? It seems that it is primarily to communicate amongst each other and facilitate the myriad human activities that are the business of life. The spoken word and the strings of words arranged to form fluid sentences, communicating meaning, are innate to us. The assimilation of language seems to be almost effortlessly 'natural' to our very existence. Is it possible to extend this painlessly smooth acquisition to the art of understanding patterns and numbers or, more formally, to mathematics? In other words, does mathematics have a hidden language?

Why does the number of petals on a flower typically equal five? We are rarely ever asked to count petals and leaves on a stem when we first learn the basic numbers. Later, when we encounter the number pi, or 3.1415..., we are merely told that it is the ratio of the circumference and diameter of any circle. We are never told the magical story of the average geological river – that the 'string length' divided by the length as the 'crow flies' from the start to the end of the river approximates pi.

Possibly, the very first time we get introduced to numbers and to the subject of 'mathematics' is when we first go to school or prior to that by our parents in preparation for school. By the time we arrive at school, however, it seems it is already too late to experience that sense of ease or 'effortlessness' in the acquisition of mathematical skill. With the introduction of a simple trick or an observation about a routine process, the child might light up with his imagination soaring. For instance, the 11 times tables is often a child's favourite because of its neat symmetry:

11x1=11, 11x2=22, 11x3=33, and so forth. But there is a further delightful property that can be attributed to 11. Any number, say 23, multiplied by 11, can be written down swiftly by noticing:

23 x 11 = 2_3 with the gap filled by the sum of the digits 2 and 3, which would be 5 in this case.

Thus, 23x11 = 253, where the middle digit is just the sum of the first number 2 and the next number 3.

Before we go any further, it may be wise to start at the very beginning and to ask if mathematics even has a language and what would that mean if it did? Further questions emerge from this query: is the basis of our mathematical understanding of existence linked in any way to patterns and arrangements that occur in nature? Do animals have a sense of order, a sense of 'counting?' More directly, do animals actually count and do they find it necessary to have a sense of 'quantity' for their own survival and thriving? When I was a small child, I wondered if birds could count. What if we removed her eggs one at a time, would she get agitated? At what point would she start panicking? And would that agitation arise because she knows exactly how many eggs have been lost to the thief, or is it a general sense of reduced 'quantity' that alerts her to tragic danger?

What about animals who often have to assess length and distance in deciding whether to jump, like in the case of an agile cat from a high ledge? What about the situation where the animal needs to make a quick calculation whether to choose 'fight or flight' when it comes to assessing the number of their adversaries, were they to show up? We may recall watching nature programmes on TV where a large animal like a giraffe is attacked by a group of hyenas. Often the outcome of the attack is hard to determine beforehand and could go either way. What should the policy of the giraffe be – how to assess strength and number?

The drift of the above questions and examples is to explore the idea that quantity, number, and patterns, form the basis of our understanding of the universe. If animals, plants and other myriad creatures use a sense of quantity to navigate their lives it very well may be that we as humans also have embedded within ourselves a more natural sense of mathematics and number. At this juncture, we may also question how we are first introduced to the world of 'quantity' and to the assessment of 'less' and 'more,' of 'equal,' 'true' and 'false.' Were we to re-imagine our educational system which we are typically handed down, automatically, through the schooling policies and curriculum, how would we better present ideas about mathematics so that the subject emerges more allied to nature? And also through a different approach to learning mathematics, mainly through examples within nature and experience, could we help create a sense of wonderment and curiosity in the young learner? It is time to delve into new ways of exploration, rather than persist with a mechanical introduction to the subject that oftentimes leads to children losing interest in mathematics early on in their educational trajectories.

The term mathematics derives from the Greek *mathematikos* which means that quality of things that are already given and yet must be remembered. By the very root of the word, the nature of mathematics is pre-given. Usually, by the time children have completed their primary education, many of them are already tuned away from mathematics. One of the reasons might be the lack of connectivity of the subject to anything else outside the classroom. Were they to get a sense of rhythm, music, art, structure, architecture and beauty, through their study of mathematics, a revolution in education could occur before our very eyes.

From the earliest times it appears that mathematics arose in the oldest societies from the need to count and tally numbers, mostly through matching objects to knots, stones or inscriptions on wood. It may be interesting to ask what kind of relationship to mathematics would emerge and develop merely from noticing patterns, even in the arrangements of numbers such as the 'times tables' without an urgent application or requirement for tallying and counting, such as trade. As a young student in primary school I was subjected to learning the tables by heart up to the twelve times table and I did so mechanically and diligently. It is only as an adult that I noticed the patterns that, say, the nine times table had:

9x1 = 0 9

9x2 = 1 8

9x3 = 2 7

9x4 = 3 6

9x5 = 4 5

9x6 = 5 4

9x7 = 6 3

9x8 = 7 2

9x9 = 8 1

9x10 = 9 0

The right hand side of the answers suggest an intriguing pattern where the units column starts with 9 and ends in 0 and the tens column starts at 0 and ends in 9, and the sum of the digits of the units and tens numbers in any row add up to nine. I am sure that being aware of such patterns and symmetrical structures would have transfixed me as a child, making the entire process of learning mathematics much more exciting and satisfying.

The most significant question to ask is: if the numbers that we encounter in school that start from natural numbers, 1,2,3,4... to whole numbers with the introduction of the number 0, meaning 0,1,2,3,4…, are somehow intrinsic to the human instinct? On the other hand, is there some other way of organising number or quantity or pattern, that is more fundamental to our innate sense of *calculia*? The process of the acquisition of language from the time of birth involves the arrangement in our minds of the variety of sounds that we are immersed in from the outset. There is a sense of language that predates birth and therefore the new born child seems well poised to create structure and assign meaning to words and form thoughts when it encounters speech. Making sense of the apparently random sets of noises and thereafter giving meaning to these staccato sounds seems to be the forte of the infant. Is there also such a potential for picking up numeracy through merely being; through watching and observing faces, through noticing the arrangement of leaves, plants, flowers, through watching clouds, birds and birds in flight formation, through staring at the night sky with its tapestry of stars, and through the infinite variety of stimuli present in the infant's natural environment? Is it clear to the child even without being told in a classroom that the idea of number is embedded all around us and indeed, as a start, within our own bodies, in the sets of fingers and toes and in the symmetry of our own form.

Thus, well before we meet our teachers in school, we have already encountered nature as the great teacher and have accessed the unique 'language' of mathematics even without being aware of it. If only this effortless learning and understanding and the natural curiosity that is prevalent in the young child were to be preserved for the years that follow, would learning not be a much more effortless, natural and free flowing process? One that brings great joy from the very inception of the process rather than drear and eventual reluctance in the vast majority of pupils?

The so-called elementary basics of numeracy taught in schools quickly lead to the four functions of addition, subtraction, multiplication and division and thereafter proceed to fractions and ratio. Straight away it becomes apparent to the young learner that a new type of language with its own patterns and rules and some clear cut operations are being presented to them. New symbols and a variety of sophisticated jargon are introduced over time. It is indeed unfortunate that the meta mathematics' behind the rules that are thrust upon us are never fully explained. Even the bright and curious eight year old who chances upon apparent discrepancies is sometimes blocked by the teacher who would, rather than deal with an overeager and interrupting, pesky student, prefer to go on with the lesson and finish the tasks of the day.

I once had a very intriguing and insightful exchange with one of the professors of my university who mentioned an anecdote from his early years as a student. He related a story from his childhood in his mathematics class in Russia, and his many encounters with his class teacher. During one such exchange, he remembers his teacher warning him about 'named' numbers. For instance, she cautioned him against adding apples to people. After some deliberation, he asked why in that case can we divide apples by people?

10 apples: 5 people = 2 apples.

What is more discomforting, when we distribute 10 apples giving 2 apples to a person each, we have,

10 apples: 2 apples = 5 people.

Where do 'people' on the right-hand side of the equation come from? Why do 'people' appear and not, say, 'kids'? There were no 'people' on the left-hand side of the operation. How do numbers on the left-hand side know the name of the number on the right-hand side? In this case the teacher failed to outline the fact that in the first example 'apples' divided by 'people' should give 'apples per people' and in the second case 'apples' is divided by 'apples per people' to give 'people' on the right hand side of the equation.

Were there to be a natural language of mathematics that is far more fluid and easy to grasp than has been historically presented, what would the natural precursor to this language be? Would the order in nature, and the order in chaos be a part of this great understanding? A few years ago, I embarked upon an experiment. A friend of mine had requested that I teach his daughter for a month or two during a trip to visit him at his farm in south India. His daughter, he woefully reported, was not too keen on studying mathematics and had a hard time learning the subject in school. Taking this up as a challenge I quickly outlined a rough curriculum for that short time period – one that would be a joint journey of discovery as though no one knew the rules of mathematics but derived them from scratch upon encountering them naturally in the environment. The emergence of this new kind of mathematics from everyday experiences in living would shine more light on the technical formulations of the lessons that were so 'boring' in the classroom and put them in a broader, more dynamic context of real experiences. Our classroom was the large farm forest outside and our learning came from acutely observing all that went on in the sprawling gardens. At the end of the course, my friend's daughter was determined to become a mathematician!

This book at the very core is trying to say something very simple. It is an attempt to recognise the effortlessness with which learning can take place if it emerges naturally, much like one's mother tongue. One's own mother tongue can be *felt*, one does not actively have to acquire it. Nonetheless, when one grows up and decides to learn a foreign language, that acquisition feels much more laboured and requires substantial efforts in adjusting to new sounds of syllables, new structure and vocabulary. With effort and perseverance, of course, one can master even a foreign tongue. It is true, though, that some are daunted by the process and stop their endeavours mid-way. What if the foreign language seemed like one's mother tongue and a natural affinity was felt from the very beginning – would one not find the entire process of learning something new utterly welcoming and pleasurable? What if there is a window into the learning of mathematics such that it begins to feel like one's native tongue, or at least much more enjoyable and approachable, full of magic and wonderment?

What are the first things that the infant experiences while in the womb? All kinds of sounds and vibrations, light and shadows, conversations, and possibly a general sense of the environment. There is no separate 'attempt' to pick up language. It is all a holistic part of the development of the child both in the prenatal state and after birth. Is it a stretch of the imagination to believe that the very same experiences in the womb and shortly after birth also lead to an understanding and assimilation of the 'mathematical' world of patterns outside? Therefore, the visual world at large that is also perceived via sounds begins to teach us from the very moment we are born. From the song of the koel bird to the shadows of leaves on a wall at night, our very act of being and our total engagement with the environment and our constant observations begin to teach us the fundamental structure of mathematics.

As outlined in the previous examples, some of the many hiccups children experience early on in the classroom relate to the very early experiences of mathematics in a formal context. One example is when they first encounter the order equivalence in addition: 2+3 is 3+2. Why is that? What is it about numbers that allow this equivalence. That and many other rules that are thrust upon children are typically never discussed by the class teacher. It is remarkable how profound mathematics can be even at the elementary school level where the pupils are introduced to addition, subtraction, multiplication, division, fractions and decimals. Those very operations and concepts, without delving into more involved contexts, can open up a unique horizon full of fascinating secrets and the promise of further discovery. The profound meta structures underlying the simplest of mathematical principles are often picked up by the keen and observant student, but unfortunately, there is no room to pursue a simple thought either because of the large class sizes or a lack of enthusiasm on the teacher's part.

Shortly after the elementary operations are introduced in school, many pupils are 'lost' to mathematics when they encounter a mixing of number and alphabet as in the case of algebra. It is indeed sad that the vast beauty of mathematics that can be discovered without much more sophistication than the elementary operations and definitions, are potentially permanently lost when the young student secretly decides to dislike mathematics. Frequently, this is reinforced by the parents via numerous channels; pressure tactics and fear being a typical instrument that is regularly used, much to the detriment of the child's flowering.

One can imagine, if the entire world of the introduction to mathematics was turned on its head, and shortly after the introduction of numbers, a systematic building of mathematics as it occurs in ordinary life experience was implemented, we might chance upon a more revolutionary type of learning. One that leads to extraordinary effects on society itself. This would imply a complete re-imagination of numerical education. A joint discovery, through the teacher and pupil relationship, of the very many hidden secrets of mathematics that lie right before us, waiting to be explored.

Another flaw in the conventional 'system' is that the philosophical basis of numbers or what one might call the theology of arithmetic is seldom introduced to the child. Hence, the fundamental philosophical nature of '1' as unity, as encompassing everything, the totality of everything, is never presented clearly. It might be argued that the number '3' is the first true number, where a sense of 'the many' is demonstrated, as opposed to the concept of oneness in the case of unity, and the concept of two or duality in the case of the numeral '2.' The importance of the numbers 1 to 9 and the introduction of '0' as an important place holder is never fundamentally clear to the early learner. They are taught instead to accept the rules as they are rolled out via the primary curriculum or syllabus, with little emphasis on creative questioning or questing into the basis of many of these rules and memorisations. Children are fascinated with the quality of numbers more than quantity whereas the system is geared towards quantity. The emphasis on the quality of number as opposed to quantity might be a smoother way to introduce mathematics as an interesting but worthy experiment.

As a child, it took me years to understand that addition and multiplication were related – and that multiplication was merely a way of grouping numbers and organising their repeated addition. Instead, I took the multiplication tables as gospel truth to be committed to memory and mastered for posterity.

The versatile and highly useful number 'pi' showed up much later in my long road to mathematical education. I did not hear of it for the longest time, and certainly not until I saw it in the context of mensuration and circles. Instead, I had learned about decimals early on, quite out of any context and plainly as a rule to be memorised and vaguely understood as a natural progression of chopping things up into tens. What if our introduction to decimals happens through the introduction of circles – in context – with the child being reintroduced to regular shaped concentric bodies like sections of fruit, leading to the discovery of 'pi'? This might be in sequence to the child's notion of 'round' through observing spheroidal fruits and vegetables – and a sense of irregularity upon noticing the occasional odd looking potato.

Surely the young student, in the days of their infancy, had noticed a round moon and other circular shapes. Therefore, from the very early years one could introduce the idea of a circle and calculate its attributes like the diameters and circumferences of different sized rings. Shortly after the ideas of division and decimals are introduced, our young learners may be asked to explore how the ratio pi naturally arises out of the circumference and diameter relationship. A way to whet the appetite of young children would be to encourage them to question everything they see from the start. A good question to reflect on, in the context of the number pi, is the following: Is pi at all a number if it appears to have never ending decimal places seemingly spreading to infinity? How can this number never end? And if it never ends, can the 'next' number that is not pi, even begin?

If mathematics followed a more natural pathway of human experience, full of texture and shapes, both regular and amorphous, we could hope for a smoother journey towards our numerical education. The simple act of observing objects that occur both around our houses and the surroundings: trees, plants, seeds and insects, the 'heavens and the earths and everything in between,' would guide us towards the grasping of deep ideas like discrete and continuous, the infinitesimal and the vast, zero and infinity. Even the acquisition of new symbols and representations would presumably be easier to assimilate within our minds. Moreover, an exposure to the history of calculation and numerical tabulation in the context of trade, astronomy, astrology, farming and other timeless activities that we have indulged in as a human race would intertwine history with the progression of the human story of mathematics. This may eventually lead to a more friendly development of the 'language of mathematics' such that nature's mother tongue might be grasped more readily.

CHAPTER TWO

NATURE AS THE FIRST CLASSROOM

"Biologists say that trees are social beings. They can count, learn and remember. They nurse sick neighbours, warn each other of danger by sending electrical signals across a fungal network known as the 'wood wide web' and for reasons unknown, keep the ancient stumps of long felled companions alive for centuries by feeding them a sugar solution through their roots."

—Mary Snowden

In Scandinavian countries like Denmark and Sweden there is an option for kindergarten children to be taken to a 'forest school,' not merely for the occasional excursion but for 'school' every day. The forest school became popular in Scandinavia in the 1980's as a response to limited indoor space for preschool activities. It was believed, and later demonstrated that the 'outdoor' nature of schooling was of tremendous benefit to the overall development of the young children.

For those teachers and facilitators that conducted the forest schools, with their rising popularity and appeal, it became evident that the children were able to seamlessly integrate the natural resources around them into their play and therefore into their learning. The children, through their interactions with their natural environment, developed a greater sense of curiosity and imagination and showed better self-control and mutual respect. They also demonstrated better resilience and were truly able to flourish in this environment of leisure and gleeful play. From all accounts, these forest schools met the true import of the root meaning of 'school.' Ironically, the word 'school' is derived from the Latin '*schola*' meaning 'leisure for learning,' or the more drastic Greek word '*skhole*' meaning 'leisure, spare time, rest and ease.'

About fifteen years ago a friend started a forest kindergarten preschool in Hampshire, UK, based on the Scandinavian model. Over time her school flourished as the children learned in the natural setting of a local forest and in a neighbouring barn where they would gather for meals and activities. Over the months of the year, the young learners adapted to seasonal changes in the environment – from the warm days of the summer to the muggy and rainy days of the intermittent wet seasons, to the chilly and snow laden days of the winter months. The children were mainly learning through free play and social interactions as this type of 'early years education' was informal and relaxed and created an atmosphere of 'home.' Being educated without the pressures of league tables must have had a positive impact, as the response from the parents community and even the educational assessment authorities was extremely heartening, with outstanding assessment scores obtained yearly. The children's engagement with life remains the central philosophy of the teaching model in such forest schools, as typically the children play with what they find and also develop ways of self-engagement and entertainment. They all learn to collectively play, gather to share meals and socially interact with each other, thereby building their own community.

One would also expect there to be a latent learning of mathematical principles as the children are left to find play-objects and resources on their own, allowing for the development of greater creativity and enterprise. The very many natural sights that are bestowed on these young learners through their forest experiences are a form of hidden instruction, carrying mathematical meaning. Balancing on logs of wood, or gazing at the clouds and birds in the sky offer many delightful insights into the physics of natural phenomena. In contrast, students in boxed classrooms seldom get a chance to watch the immense palette of colours showed off by the vast tapestry of nature. Indeed, something as seemingly mundane as cloud formation – from serrations to large puffy bulbous types of cumulus clouds – lead to much wonderment, and is certainly a special example of mathematical order and design.

Our own integration with nature and the many secrets it has to offer may have been more *ad hoc,* as many of us may not have attended a school within a natural setting. Even so, this does not exclude us from making our own observations and scientific experiments. Upon taking a walk along the banks of a river, we might have noticed ripples on the sand – one of nature's most spectacular examples of self-organisation. These patterns develop from the dynamics of individual grains in motion and provide a perfect template to contemplate upon and admire nature's 'blackboard' as it teaches us about patterns and outward order. We may remember dropping a stone into a pond or river and watching the ripples spread concentrically across as they merged into the more distant, calmer waters. No one points out the properties of ripple formation or the curious nature of water and water waves. These are left to the eager and able mathematics student to be addressed only if they choose to pursue fluid dynamics as a specialised subject in the distant future. Mentioning the interesting features of ripple formation, including replacing the stone with a cork and watching it bob up and down, might indeed gratify the inquisitive child. Splashing stones into the pond's still surface generates many other noteworthy features like bubbles along with ripples. These bubbles and also the 'crowns' of rainwater splashes are additional instances where, inadvertently, many mathematical features such as the transitory nature of bubble formation are noted by the young student.

Long before the young mathematics student is introduced to the word 'fractal,' she has already noticed the repeated, self-similar patterns, in the branches of trees, in ferns and even in snowflakes. The fascinating study of 'repeated patterns' in nature is undertaken by the mindful child merely by *observation*. It seems when children are given space and opportunity, they automatically begin to learn from their environment. They become keenly observant and perceptive, lapping up the treasures that nature places before them – eager to make sense of the various sounds, smells, structures, and textures they are exposed to.

Fractal tree branches

Children who have not been explicitly taught the numerals except through picking up some numbers via speech might learn the idea of *quantity* and *number* from playing with objects like stones. For the first balancing act, the child gathers some stones of various sizes, piling them up with the largest ones at the bottom of the stack. The concept of stable, unstable or neutral equilibrium is learned instinctively without any intermediary class teacher. The mere act of gathering stones and setting them up as a stack or a conical collection gives a sense of both number and balance.

Another activity that might also present a basic understanding of 'mathematical packing' would be to notice how sand particles and pebbles of different sizes distribute themselves within a container. Do the smaller particles fit in between the larger stones or grains? If so, how would the distribution occur? Without attempting to explicitly answer these questions, the child gets a sense of packing order. The study of the flow of grains may continue at the kitchen table later on when the child notices the parent pouring flour or polenta into a container, forming a continuous stream.

Access to either one's own garden or playing areas in the general commons is also highly desirable for the overall learning experience. Seasonally, autumn leaves arrange themselves into a beautiful carpet of amber and yellow hue. Picking up a leaf, especially one with 'corners' like the maple leaf and counting their vertices may well be a joyful activity for the young student of early numeracy. As they grow older, merely counting leaves or pebbles could give way to questioning why it is rare to find even numbered leaves – the ever elusive four leaved clover? Nevertheless, it is not uncommon to find eight-petalled flowers.

Nature also throws up a variety of puzzles to be noticed by the alert mathematics student. For example, why do certain insect species like cicadas only appear in years that are a prime number away from the previous, or prime numbered, year intervals? A biologists way of understanding this unique phenomena is so that these cicadas are relatively immune to the shorter life cycles of their predators. As an example, cicadas that emerge after every seventeen years will only encounter their predator (who might have a five-year cycle) every eighty-five years, giving them an advantage over other subspecies that might be less adapted to this protective mechanism. Even though the life cycles of cicadas and their link to prime numbers may escape the young student, they are sure to notice other cycles like the periodic flowering of trees and ripening of fruit or the practically precise arrival of the seasons. The most obvious of the daily cycles of nature is the emergence of the sun and the moon. They are indicators that some things in life seem to have a periodic guarantee with a soothing consistency.

Waking up to the patterns of natural sounds and listening to early morning bird call brings a serene feeling of well being and a sense that the day is ready for unfolding. Depending on where we live, the variety of different bird songs can themselves create an early morning symphony. Writers, poets and lyricists might compare the singing of birds to the music composed and conducted by humans. A study of bird song in the Proceedings of the National Academy of Sciences documented the pitches of 114 different song types from 14 male hermit thrushes. After isolating the frequencies corresponding to each note and calculating the mathematical relationships between the pitches, the authors were pleasantly surprised to discover that the notes in the bird's songs fitted into the harmonic series which are a sequence of notes based on multiples of a baseline pitch. The harmonic series are a fundamental part of human music as it helps ascertain what pitch combinations are aesthetically pleasing to the ear.

We may proceed to surmise that we recognise in bird song a sense of the familiar and pleasant and even go on to reproduce some of these harmonics in our own music. The mathematical patterns that are produced in the natural world through these frequencies of bird calls resonate with us in some way and are picked up and noted by us from the earliest exposures to them. Is nature teaching us harmonies through its infinite variety of sounds and vibrations? Is our composed music thus a composite of these infinite varieties of exposures and experiences from within the natural world?

Witnessing the beauty of our natural world surrounding us is an indelible part of the human experience. Most fascinating of all is the extraordinary effect of first noticing a rainbow with its myriad colours painted across the sky. The more vivid and spectacular it is, the more etched in our memories it becomes. We look for rainbows in many contexts, one of them being in a prism splitting rays of sunlight into an assembly of colours. What does the young learner pick up from noticing this spectacular magical feat of nature? The fact that so many colours exist like ordered ribbons of colourful light is itself a puzzlement. The same effect is felt when oil mixes in a puddle to create a thin film of mixed colours. Even though the callow learner may not have encountered the concept of thin film interference in physics, the sight of these rainbow like colours mixing and forming in a dirty puddle brings an element of surprise and an eagerness to persist with questioning the incredible diversity of natural occurrences.

Gardening and farming is another sphere where one might gain immense insights into the underlying processes of soil-plant interactions that govern forest growth and even agriculture. Ideas of discrete and continuous regarding the multifarious nature of soil itself, and the nebulous zone in which soil ends and plant or growing organic matter begins, are all presented to the inquisitive learner granting numerous insights into the nature of matter. Sometimes loose soil exposes the root system of a bulbous plant that has overturned to reveal the network of roots and root hairs. The observer notes the tiny ecosystem of soil-plant-roots in delightful contentment, by merely witnessing this fascinating mini-terrain. I am sure no calculations are going on in their heads about the 'microbial dynamics in the vicinity of a single root' but indirectly they are learning exactly that!

Another example of the great cosmos teaching us tremendous secrets and revealing numerous mysterious unknowns, is the night sky. By merely watching the sky on a clear night, so many stars spanning the vast unknown begin to suggest that there is much that will remain hidden from us no matter how much we know. The fact that the entire sky with its myriad features can 'fit' into our retina in a tiny image itself is a miracle. According to David Bohm (1917-1992), the well-known cosmologist and physicist, the implicate (or hidden) order is the ground from which reality emerges. Thus, the sky in all its splendid glory is revealing the hidden order of things merely by being witnessed by us.

The moon is perhaps the most romantic object above, inspiring many a poem. Even a small child will notice how it seems to 'walk' with us wherever we go. The dark spotty features on the moon are surely the markings of the face of the man on the moon, or they are merely interesting geological features that appear like markings from far away.

Simple forms of the laws of Newton are also inferred by the young student before they encounter them officially in any formal context. The second law is specifically noticed when nearly all objects, except the fascinating helium balloon that drifts up instead of down, fall downwards when released from a height. The young child has received a 'demonstration' of the gravitational pull of the earth when he or she has tried to copy a bird and lifted off from a stump or high table, inadvertently falling to the ground after a failed experiment. Parabolic trajectories of projectiles aimed at adults or fellow children often miss their mark especially if the child is surprised to see that the target was missed because the aimed object followed a 'parabolic' route downwards and did not continue in a straight line. These refreshingly precious lessons from unstructured play are possibly some of the deepest mathematical and physical insights that help to guide the child into adulthood.

When the child encounters the colourful beetle with its folded symmetry or a gorgeous butterfly with unfolded wings flying away, it has learned a few things about not only the symmetries of mathematics but beauty and art as well. Along with the active beetle the child may also notice an ant colony with several busy bodies carrying food particles several sizes bigger than its own body, marching in order. How can they possibly carry out such an impossible feat? This question is fundamental to science and in fact lays the foundation of Galileo Galilei's (1564-1642) lecture 'Two New Sciences,' where he talks about his investigations on the mechanics of scaling. In one of his lectures that finally gave him a university teaching position, Galileo asserts that 'the surface of a small solid is comparatively greater than that of a large one.' The child has already noted this physical reality by his curious response to the marching ants with their 'heavy loads' and a later encounter with an elephant if he or she were to ever meet one. In 'scaling' up because of its size, the elephant loses some elegance in forming its gigantic but non-regular body.

Beside the soothing quality of the natural sounds of the forest, the quality of air, sound, light, and soil, the myriad vibrations found in the natural environment must surely serve in honing the cognitive and analytical skills of each individual child. Thus, foundational mathematics, in the true sense of the word 'foundation,' must begin with our relationship with nature.

CHAPTER THREE

PATTERNS IN NATURE

"What we call chaos is just patterns we haven't recognized. What we call random is just patterns we can't decipher."

—Chuck Palahniuk

Nature sings the song of mathematics. Anyone who has spent time in a flourishing garden has noted the variegated patterns in the colourful flowers and the dancing bees flirting with them as they visit flower to flower. But most noteworthy of all might have been the discovery that the number of petals in a flower are typically five or eight. The odd discoverer may even notice a pattern by realising that the petals almost always stick to certain numbers like 5, 8, 13, 21 or 34. Where do these numbers emerge from? And why don't other numbers like, say, four (the famous rare four-leaved clover) or seven ordinarily not appear?

These patterns that show up in nature, within the spiral symmetry of a sunflower or the arrangements of leaves on a stem or in the spirals of an acorn, all arise out of the same underlying mathematics that appears repeatedly in numerous other instances in the natural environment. We encounter these very symmetries in fruits like the pineapple or apple; we can count the knots on a pineapple which typically are 8, 13, 21 or 34. If you cut an apple across in half, the seeds within happen to be 5 in number. Again, we meet the same repeated pattern through the number 5. These numbers could be identified as the famous Fibonacci series of numbers that start 1,1, 2, 3, 5, 8, 13, 21, 34, 55... *ad infinitum*, and are composed by merely adding the previous two numbers starting from 0 and 1. Thus, 0 added to 1 is 1, and 1 added to the previous 1 is 2 and 1 added to 2 is 3, thereby forming the early numbers of this infinite series. We shall look into the Fibonacci series in depth in Chapter 6.

If we replaced the classroom learning of mathematics by a more natural manner in which it is discovered, such as through spending time in nature and watching with great alertness what patterns present themselves, we might chance upon a different kind of mathematical curriculum. One that is more organic and arises out of participation in nature, without effort, and is suggested by the human experience and falls within the timeline of our first encounters with these natural phenomena. Our learning of number and the four main operations of addition, subtraction, multiplication and division can be concurrent with the experience of nature in its additive or multiplicative splendour.

From the very outset we are immersed in a sea of mathematical patterns. Nowhere is it more stark than when we notice the scales of sea animals, their shells forming extraordinary spirals. It is also interesting to observe line symmetry in seashells, crabs, star fish and in innumerable varieties of sea creatures that make it to the beach, displaying their amazing arrays of patterns. The artistic ripples on the scales of angel fish and the diverse types of 'etchings' on the bodies of clown trigger fish are nature's artwork that also appear to have a mathematical template or rule book.

Patterns occur in nature in nearly every context, in water bodies like ponds, rivers, backwater enclaves of bays and seas, in ocean waves and in rainwater puddles. The interesting mathematical relationships between the speed, wavelength and depth of waves govern how they travel or behave. The order in the seemingly chaotic panoply of nature's offerings would capture any artistic mind. An early exposure and study of these numerous natural phenomena may indeed form an indelible blueprint in the minds of the young learners, such that the future study of these very images and concepts are more graspable and happen in a stress free manner leading to a better fundamental understanding of the topics and subject of study.

There is hidden mathematical order in the gathering of flocks of birds, their pattern formation as they migrate, and among crowds of people as they assemble. Many patterns are frequently spotted by children, who are filled with awe and admiration when they first discover these ordered collections. Perhaps the deep rooted mathematical patterns and an innate comprehension of these designs are already encased within us only to be discovered later. It is also believed by some that mathematics is an artificial interpretation of our surroundings and the world, one that works in the large part and seems to have brought us technology and the technological world we inhabit. However, there are still many occurrences in nature such as the prediction of weather or even the anticipation of earthquakes that don't seem to offer themselves so easily to mathematical analysis. Whatever the case may be, perhaps the truth lies in a combination of these two disparate views that mathematics is innate or is artificial. By exposing early learners to a greater gamut of 'nature' experiences, through trundling along a forest path or by spending endless hours on a beach or by just observing the sky as the first stars start twinkling in the darkness, some of these mathematical secrets will be revealed and automatically assimilated. Akin to a native language.

Patterns, as they are formed in nature, are sometimes born of efficiency. The shortest path, the tightest fit or the nearest neighbour are all instances of this striving for efficiency – even a minimisation of 'energy.' The technical term for this kind of efficient packing is called 'the voronoi pattern.' Mathematically speaking, each cell in a voronoi pattern has a seed point and everything that is inside a cell is closer to it than to any other seed. The lines between cells are halfway between neighbouring seeds. Eating one's favourite corn on the cob meal or indeed chomping on a head of garlic alerts one to this interesting cellular behaviour.

Another fascinating example of patterns in nature that commonly appear in the world of bees is the hexagonal structure of the honeycomb. In 1999 'The honeycomb conjecture' was proved by the mathematician Thomas Hales, that states that a hexagonal grid which is the honeycomb structure, is the best way to divide a surface into regions of equal area with the least total perimeter. Nature's patterns, as beautiful as they look, are also among the most efficient it seems. Thrift, efficiency and efficacy are the guiding principles that determine the type of mathematical forms that govern the particular phenomena being examined. Another example of these honeycomb patterns happen to be in weathered bedrock and consist of tightly adjoining patterned cavities.

Spider webs are highly optimized and lightweight structures with incredibly ordered radially or spirally diverging 'silk' threads forming the overall structure. These webs are also extremely adaptable to external forces such as rain or wind. Another extraordinary property of the spider web is that the force distribution remains unchanged when any spiral thread is broken. These patterns in nature, such as spider webs, are highly instructive to us for studying physical properties like tensile strength and examining the distribution of forces in radial patterns. Another insightful study is that of the chemical properties of 'spider's silk' as a high performance fiber. Radially emanating and spiral structures are not merely found among spiders and other insects building their home, but within galaxies that comprise innumerable stars and galactic matter, all of which are spread over the entire cosmos and also have unique and predictable mathematical formulations.

Fractal patterns that create apparently self-similar designs are all around us, forming repeating sets at different scales as they grow. They arise in snowflakes which have six-fold symmetry, in trees as they branch out, in lightning patterns in the sky, in the garden fern and seem to be around us everywhere in nature. Although these self-repeating patterns provide a stunning vista before our eyes, they typically do not repeat endlessly and break down at some point unlike a mathematically repeated pattern such as the Mandelbrot set that does go on forever. Nevertheless, the plant kingdom has a vast repertoire of fractal patterns that form from the ground up – when the sprout emerges from the ground and splits into branches, and each of these split further. The repetition of branching that forms the tree generates the tree's self-similarity.

Spirals are also commonly observed as growing patterns, for instance in the nautilus shell of the cephalopod mollusc where each chamber of the shell is an approximate copy of the next one, scaled by a factor and arranged in a logarithmic spiral. This example is also a special case of the self-similarity repeated pattern discussed earlier. Children playing around in the garden or grown ups taking a nature walk through a forest path will undoubtedly come across numerous instances of these self-similar designs in their environment – composite flower heads, seed heads like those of the sunflower or daisy and numerous other dynamic systems that alert them to the enormous corpus of nature's splendid bounty. Returning home to notice multiple 'Fibonacci spirals' in a cross-section of the red cabbage being prepared for lunch would certainly add to the sense of wonder we all experience. Spirals not only appear terrestrially but in the vast sky above. Most noteworthy among them may be the countless galaxies that all display rotational 'swirling' behaviour.

Galaxies present among the most beautiful spiral patterns and are now understood to form mainly as a result of gravitational forces between stars and gases in a shearing, rotating disk. In other words the joint motions and gravitational effects of all the stars create the spectacular patterns of the rotating galactic structure. The spiral arms of the galaxies are waves that move through the gas and stars in the disk. The complexity of studying how the spiral arms in the galaxies exactly emerge is a matter of ongoing study. This is another example of elaborate but beautiful patterns in nature that appear ubiquitously, including in inter-stellar space.

A further example of patterns that commonly show up around us are the spots and stripes that can be seen on animal skin. The patterns that appear as zebra stripes or those on the mighty Bengal tiger, or indeed the spots on leopards, were predicted by Alan Turing in 1952 when he described what he called reaction-diffusion equations in his seminal paper 'The chemical basis of morphogenesis.' He was so ahead of his time in laying the mathematical foundations for the study of the emergence of patterns that we are still baffled by the sophistication of this subject to this day. Apparently, patterns could spontaneously appear using a protein's natural tendency to move randomly through tissue. This interaction with other cells and proteins lead to the vast array of designs we observe around us in nature. Even without knowing about any chemical or cellular basis for these patterns, the observant onlooker is fascinated by the stunning display of a peacock showing off its radial and colourful feathers, one of the many instances of spectacular exhibition we encounter in nature. The more we are exposed to a natural environment where we can witness such a demonstration of patterns, the more will we understand how patterns serve to camouflage or attract, deceive a predator or display to an admirer. Sometimes, patterns in nature may seem to serve no purpose or have no witnesses – even in these instances they all collude to participate in the grand collective dance of events that comprise our beautiful cosmos.

The chaotic order in natural phenomena is captured via the study of 'chaos theory' that helps us understand many patterns in nature. It states that within the apparent randomness of chaotic complex systems there are underlying patterns and there is self-similarity generation. We notice these replications in fractals and other self-organisational behaviors manifested in repetitive cycles.

The patterns we see in fluid flow meanderings, in vortex sheets, in waves and dunes and in the propagation of cracks are fascinating subjects in themselves and mathematical models frequently try to predict the seemingly random creation of patterns. In 1982 Benoit Mandelbrot published 'The Fractal Geometry of Nature' which became a classic work in chaos theory.

It has been close to 500 years since Leonardo da Vinci performed the first scientific examination of the human body in Florence. Among his discoveries was the fact that the heart was a muscle, responsible for pumping blood through the arteries and then the veins, and was not, as was previously thought, the 'vital spirit' of the body. It is the arteries and veins, and the flow of blood within them, that primarily motivated my own doctoral studies on the mathematics of blood flow in the body. I was not surprised to observe, mathematically, the undulatory nature of arterial blood flow when encountering a constriction within the blood vessel. My numerical calculations were a mathematical simulation, using modern techniques, of physiological flow like that arising in blood vessels within the body encountering a constriction. The constriction mimicked an arterial blockage or an aneurysm. The results indicated complex 'mathematical' oscillations in the flow, with different types of periodicity, depending on the response of the visco-elastic walls that were modeled by a flexible tube containing the fluid (blood).

Leonardo da Vinci not only gave us tremendous insights into the processes within the human body but was also responsible for discovering that trees, as they shed their foliage, reveal a startlingly simple relationship between the size of a tree trunk and its branches. He scribbled in his notebook the following observation, "all the branches of a tree at every stage of its height when together are equal in thickness to the trunk." These trees are obviously fractal in structure, meaning that the patterns created by the larger structures like the main branches repeat themselves in the smaller branches. Thus, the small is a reflection of the large and vice versa.

Recent mathematical studies of biological systems have delved into areas ranging from physiological flows to population dynamics within groups. From our very early interactions we observe the behaviour of large populations of humans gathering together, along with ants, bees and other insect colonies collecting in groups. The social interactions of many creatures, along with the complex purposes of interaction, reproduction and 'society building' have an external structure too. These are also noted by the young student and the principles that underlie such collective behaviour are probably intuitively arrived upon. Collective societies, be they human or animal groups, make informed consensus decisions—how they arrive upon them has a mathematical query built into it. How do ecological interactions produce population dynamics and how does information flow within the individuals in the collective body in a continually changing environment? Mathematical models are set up to answer such questions by current researchers of mathematical biology. We have been continuously discovering the richness of our ecological system, of our environment, and the universe beyond, from ancient times to the present day. We may, however, have skipped the overall intuitive response of humans to many of the intangible patterns in geological phenomena such as the arrival of earthquakes. Why is it that many animals, insects, and other sea creatures survive vast geological impacts like earthquakes and tsunamis? Even the so-called primitive societies seem to collect in high ground prior to the arrival of a tsunami such as the recent devastating one in 2006 that spared the inhabitants of the Sentinel Islands as they managed to escape to safety before the onslaught of the big wave. Have we, as classroom dwellers, missed some of the more intuitive and non-mathematical understandings of the makings of natural processes? What then should we do to remedy this lapse? The answer may be to simply start early with our immersion in the natural world and reclaim some of the lost knowledge.

The connection between natural patterns and mathematics would now seem obvious if we cared to look around us.

Nature, in all its myriad forms reveals its hidden mysteries to the discerning learner. This direct communion between the teacher, nature, and the taught, us, arises effortlessly without the baggage and stress associated with ordinary classroom or coursework learning. Therefore, the many patterns seen in nature are also absorbed by us effortlessly creating a silent corpus of information which is mathematical in its nature. Will this effortless learning be the foundations of a 'native language' for mathematics? Historically, there are relatively few instances of this kind of learning that is wholly unstructured but immensely instructive and thereby useful in our paths towards understanding some of the many truths and secrets of the universe.

Nature in all its manifestations appears to nurture curiosity - dropping hints about its mysteries and urging the inquisitive mind to stretch, and further extend its boundaries and insights. Nature also urges us to notice patterns, presenting them as infinite repetitions and ordered sets. It is as though nature itself provides all the instructions, waiting for us to unveil the hidden mysteries of life. When the child is exposed to these innumerable patterns in their full manifestation, he or she absorbs these experiences. This process also serves to capture some of the mathematical order waiting to be formally discovered in the future. In order to do so one might have to rely on a toolkit of interpretations in the form of a more sophisticated mathematical language. The important thing to note here is that with this exposure to a myriad different natural processes, the child grows up to be a seeker and to ask questions. Later, the young student of mathematics may fruitfully start a journey that would serve to unravel concealed elements of reality. Sometimes the questions remain with them and there are no answers. For once, to not have answers is perfectly acceptable.

CHAPTER FOUR

NUMBER IN PATTERNS

"One of the most interesting aspects of the world is that it can be considered to be made up of patterns. A pattern is essentially an arrangement. It is characterized by the order of the elements of which it is made rather than by the intrinsic nature of these elements."

—Norbert Weiner

Numbers and patterns are often indelibly linked. We typically first notice patterns in numbers when we encounter them in the multiplication tables as small children. Beside the ten and eleven times table, we do find it relatively easy to learn the nine times table probably without realising why it seems so conducive to committing to memory. As was mentioned in the first chapter, the tables have an intrinsic beauty, especially when certain properties of the right-hand side are noted. However, one can keep noticing other patterns and properties such as the example of multiplying the right-hand side of the nine times table with the number 12345679 (omitting the number 8 when writing down the consecutive numbers). One can observe that,

12345679 x 1x9 = 09 x 12345679 = 111,111,111

12345679 x 2x9 = 18 x 12345679 = 222,222,222

12345679 x 3x9 = 27 x 12345679 = 333,333,333

12345679 x 4x9 = 36 x 12345679 = 444,444,444

12345679 x 5x9 = 45 x 12345679 = 555,555,555

12345679 x 6x9 = 54 x 12345679 = 666,666,666

12345679 x 7x9 = 63 x 12345679 = 777,777,777

12345679 x 8x9 = 72 x 12345679 = 888,888,888

12345679 x 9x9 = 81 x 12345679 = 999,999,999

12345679 x 10x9 = 90 x 12345679 = 1,11,11,11,110

It is tempting to dismiss such neat patterns as quirks of the decimal system. Nevertheless, every child notices their fingers and toes, and that there are only ten each of them, and therefore one can expect a decimal system to govern much of our modern mathematics – arising from noticing the digits in our hands and feet, and thenceforth treating them as building blocks of our counting system. At the outset, the young learner could bypass multiplying each line above by hand if they did not possess a calculator, by noticing that 9 is just one short of ten, and in the case of the first line, the answer can be obtained rather easily by multiplying 12345679 by 10 and subtracting itself to give us the outcome. In other words, 12345679 x 10 - 12345679 = 123456790 -12345679 = 111,111,111 and so on and so forth. In fact, discouraging our young pupils from relying too heavily on mechanical calculators would help them in noticing many more patterns such as the above by inspection and through using simple operations of addition, subtraction and ordinary multiplication.

Another interesting pattern emerges when we multiply the nine times table's results by the number 987654321 giving us the following:

987654321 x 1x9 = 09 x 987654321 = 0888,888,8889

987654321 x 2x9 = 18 x 987654321 = 1777,777,7778

987654321 x 3x9 = 27 x 987654321 = 2666,666,6667

987654321 x 4x9 = 36 x 987654321 = 3555,555,5556

987654321 x 5x9 = 45 x 987654321 = 44444444445

987654321 x 6x9 = 54 x 987654321 = 53333333334

987654321 x 7x9 = 63 x 987654321 = 62222222223

987654321 x 8x9 = 72 x 987654321 = 71111111112

987654321 x 9x9 = 81 x 987654321 = 80000000001

987654321 x 10x9 =90 x 987654321 = 88888888890

As arithmetic becomes easier and seems more natural, numbers will begin to feel like the equivalent of the alphabet for the construction of language – the building blocks of the language of mathematics. As language comes naturally and easily with exposure and practice, so would numeracy feel automatic and an offshoot of keen observation and experience.

Palindromic numbers read the same way back and forth, just like palindromic words or names do. An example of palindromic words are 'radar,' 'rotator' and 'madam,' which read the same in either direction, right to left or left to right.

When we look at numbers, we can construct a pattern such as those shown below on the left hand side which conveniently equal the pattern of ones multiplied together on the right-hand side.

1 = 1 x 1

1 2 1 = 11x11

1 2 3 2 1 = 111x111

1 2 3 4 3 2 1 = 1111x1111

1 2 3 4 5 4 3 2 1 = 11111x11111

1 2 3 4 5 6 5 4 3 2 1 = 111111x111111

1 2 3 4 5 6 7 6 5 4 3 2 1 = 1111111x1111111

1 2 3 4 5 6 7 8 7 6 5 4 3 2 1 = 11111111x11111111

1 2 3 4 5 6 7 8 9 8 7 6 5 4 3 2 1 = 111111111x111111111

Placing 'plus signs' between the numbers on the left hand side in the equations above also lead to interesting results. Adding individual numbers that read palindromically lead to the pattern shown below.

1=1 $=1x1=1^2$

1+2+1=2+2 $=2x2=2^2$

1+2+3+2+1=3+3+3 $=3x3=3^2$

1+2+3+4+3+2+1=4+4+4+4 $=4x4=4^2$

1+2+3+4+5+4+3+2+1=5+5+5+5+5 $=5x5=5^2$

1+2+3+4+5+6+5+4+3+2+1=6+6+6+6+6+6 $=6x6=6^2$

1+2+3+4+5+6+7+6+5+4+3+2+1=7+7+7+7+

7+7+7 $=7x7=7^2$

1+2+3+4+5+6+7+8+7+6+5+4+3+2+1=8+8+

$8+8+8+8+8+8 \qquad =8x8=8^2$

$1+2+3+4+5+6+7+8+9+8+7+6+5+4+3+2+1=$

$9+9+9+9+9+9+9+9+9 \qquad =9x9=9^2$

$1+2+3+4+5+6+7+8+9+10+9+8+7+6+5+4+$

$3+2+1=10+10+10+10+10+10+10+10+10+10 \qquad =10x10=10^2$

An unexpected link with the pattern above arises from adding a sequence of odd numbers. As we know, the first few odd numbers are 1, 3, 5, 7, 9 and so on. If we conduct the exercise of adding the first few odd numbers, by starting with the first two and then the first three and continue in this manner, we get:

$$1=1=1^2$$

$$1+3=4=2^2$$

$$1+3+5=9=3^2$$

$$1+3+5+7=16=4^2$$

$$1+3+5+7+9=25=5^2$$

$$1+3+5+7+9+11=36=6^2$$

$$1+3+5+7+9+11+13=49=7^2$$

$$1+3+5+7+9+11+13+15=64=8^2$$

$$1+3+5+7+9+11+13+15+17=81=9^2$$

$$1+3+5+7+9+11+13+15+17+19=100=10^2$$

and so on. The pattern above is identical in its results to the patterns seen in the earlier example of the palindromic additions. Finding connections between seemingly unrelated patterns is a feature that occurs frequently in both nature and mathematics. The aesthetic appeal of the subject becomes more obvious the more one encounters these tidy patterns.

Other numbers showing interesting coincidences might appear as peculiarities of our number system but then it is not all that much of a quirk since we have ten fingers and toes, so there is some intrinsic intelligence in selecting the decimal as our base system. Historically, we have known how to count possibly before the discovery of the Ishango bone record. The famous Ishango bone tool has been dated to the upper paleolithic era and appears to have tally marks carved in three columns running the length of the tool. The extraordinary discovery was made near the Ishango area close to the Semliki River in modern Uganda in 1960. The latest dating suggests that it is about 25,000 years old. More than a simple tallying tool, it appears to be a lunar phase calendar as more markings were discovered under a microscope by the paleolithic archeologist Alexander Marshak (1972). Thus, from the earliest times the tendency and need to tally and record has been a part of human civilisation and experience. Some of the earliest counting systems were based on 2, akin to the binary system we use today in the realm of computers. Other bases like 12, 20 and 60 as adopted by the Mayan Civilisation, were also used.

Modern mathematics has allotted the term 'perfect number' to any number whose factors add to itself. Therefore, the numbers 6, 28, 496, and 8128 are the first few ones that have this seemingly uncommon property. The perfect numbers that have been discovered so far have another unique property – they either end in the numbers 6 or 28 which is a curious pattern in their behavior. In fact, 28 is the only known number which can be expressed as a sum of non-negative integers, 1+2+3+4+5+6+7, a sum of first primes, 2+3+5+7+11, a sum of consecutive non primes, 1+4+6+8+9, and a sum of its factors 1+2+4+7+14.

The number six appears in numerous hexagonal patterns in nature as touched upon in previous chapters. The beehive has a beautiful hexagonal network in its honeycomb structure made of honey and wax. When the honey bee first brings the liquid droplets to the hive, they are actually spherical in shape but eventually assume a honeycomb structure in order to provide the most efficient fitting. Patterns of six are seen in numerous other instances, such as in the cracking of parched earth and rock or in the shape of rock outcrops found near cliffs or beaches. Even the patterns on dragonfly wings have gorgeous hexagonal structure that glisten with kaleidoscopic beauty.

In the first chapter we had briefly touched upon an interesting natural occurrence that had a link with numbers, one in particular – the number pi denoted by the symbol π which equals 3.14152965... *ad infinitum*. Any geological river when traced from its mouth to its end has a string length and a 'shortest length' as the crow flies. The ratio between them is an approximation of pi. Why would that be so? There is no circle involved and yet this spectacular coincidence appears. There are other instances, innumerable occasions, where apparently the answer to the mathematical question is conveniently π or closely related to it, despite there being no circular orbit involved or indeed any kind of apparent outward circularity. The presence of the number pi in a variety of occurrences serves to illustrate how observation and step by step deconstruction go hand in hand in gaining numerous mathematical and physical insights into the problem. An example of how pi can creep into unexpected places is the following infinite series:

$$1 + 1/4+ 1/9 +1/16+ 1/25... = \pi/6.$$

We may notice that the denominator of each term in the series is a square of a natural number, namely of the numbers 1,2,3,4,5...

These are not the only places where π unexpectedly shows up. Consider a simple nail of size L and take a sheet of paper with lines drawn on it with a distance 2L apart. If we simply throw the nail on the piece of paper from a height it will cross the lines a proportion of times, compared to the times it misses crossing any line. It turns out that the probability of the nail crossing the line is π with greater accuracy as the number of tries increases to a large number. Here, the reader may find it startling that the number π shows up again, despite there being no obvious link to circles or circularity.

Most astonishing of the patterns and sequences of numbers that appear not only in nature, but in music, physical structures, architecture and a variety of other places is the Fibonacci Sequence briefly touched upon in Chapter 2. Leonardo of Pisa, also called Leonardo Fibonacci was born in 1170 and was an Italian mathematician. He introduced Europe to the sequence of Fibonacci numbers which he used as an example in his book "Liber Abaci." The numbers which read as 1,1,2,3,5,8,13,21,34,55…include the number 0 these days. The application and emergence of these numbers along with their more extended history, together with the related Golden Ratio, will be covered in greater depth in later chapters.

One of the reasons for writing this book is to present the idea that prior to officially learning numbers or the decimal system, patterns are recognisable with their predictive artistry and precision. These patterns that suggest numbers or quantity can be unearthed even by an eager student when they are left to explore the great world around them. Whether or not numbers systems are arbitrary, young learners with very basic tools like addition and subtraction will be able to uncover many of the apparent mysteries that lie before our very eyes and yet are frequently missed.

By using simple 'tools' like stones, fruits, other foods, art, and the great expanse of nature, children are able to garner a keen spirit of inquiry within themselves. This sense of inquiry is not restricted to children alone – all of us have this spirit of exploration and keenness to learn and deeply question. When there is an atmosphere of no competition, which tends to happen when children are very young, the infinite variety of patterns are easily assimilated by the young mind. It may outwardly appear as though nature is minimising energy in forming structures, but there is bounty and a sense of plenitude in the natural world. When there is no sense of competition but just a sense of inquiry, one might begin to challenge the epithet that we were all taught, the concept of 'the survival of the fittest.'

Without the burden of the sense of the 'fittest' and the 'best' 'surviving,' everything actually seems to subsist and even prosper if we look closely. We begin to notice with greater joy the sky above and the continuity of the routine of the rising and setting of the sun, the tides, the predictability of the seasons and the cicadic rhythms of our sleep cycles.

These natural patterns become comforting and a source of gentle instruction. An effortless assimilation of number and pattern. The patterns in number or numbers generating patterns will be quickly picked up in this open-ended environment by the alert student. A sense of wonder and an admiration for the open ended questions that arise from exploration are very important in order for the learner to embark on their path of gaining deep mathematical insights. From the patterns relating to numbers, whether they arise in tabular form through mathematical operations, or in structural arrangement in the cracks of the San Andreas fault, the learner is able to piece together a larger picture and obtain a sense of the overall structure and mapping of the universe. With the exposure to numbers and patterns in numbers, the student is able to ultimately develop an intuition which we can call 'mathematical intuition' and through this process becomes open to further discovery which is the very joy of learning and the gradual acquisition of the 'language of mathematics.'

CHAPTER FIVE

NUMBER AND THE GEOMETRY OF NATURAL FORM

"The body is a device to calculate the astronomy of the spirit. Look through that astrolabe and become oceanic.

—Rumi

The subject of anatomy has always posed a fascinating array of riddles and questions, both mathematical and physiological, to the curious examiner. These riddles and mysteries have provided sufficient fodder to justify the dissection, examination and measurement of the external and internal elements of the human body. Children frequently examine their own bodies, and those of their siblings or playmates in order to make sense of the miraculous musculo-skeletal framework they find themselves caged in. The process of the naming of the various body parts also introduces them to concepts of symmetry, beauty, order and proportion. Soon they are alerted to the fact that indeed they have one nose, one naval, one mouth and one groin along a vertical imaginary line or the sagittal plane; they also have pairs of eyes, eyebrows, ears, mammary glands and limbs. Each element of the latter 'twin' body parts has an apparently equidistant placement from the central imaginary dividing line. Hence, the human body appears to have a proportionate and identical right side and left side. As young children we do not yet capture how this seemingly innocent inspection of the body symmetry of bilateral animals including humans, point to aspects of 'Group Theory' which is a topic reserved for higher mathematics.

Radial symmetry in some sea animals like the starfish or sea anemone preserve certain aspects of symmetry such as rotation, translation and reflection – concepts that are extant in Group Theory. Even without knowing the principles of this subject, many elements of rotational invariance are encountered by merely observing animals, insects, and even our own bodies. We meet several of these incredible natural structures everyday, including those that are inanimate, in the very business of life.

Children also tend to inspect their teeth and notice their varying shapes and sizes as the 'milk' ones fall out to reveal a new set. When all the milk teeth have been replaced, we may notice a sense of symmetry in the complete set of teeth – the front two incisor teeth form a golden rectangle, with a phi ratio in the height to the width. Again, phi occurs in the ratio of the width of the first tooth to the second tooth from the centre.

The presence of flexible digits have a unique functionality and help us in holding and grabbing. It is a great revelation to the child when they discover the versatility of their hands. The dexterity of the human hand and the agility and nimbleness of human fingers have allowed us to go on to build spectacular civilizations of arresting beauty and complexity through millennia. It is no coincidence then, that among the first aspects of our hands we notice is how our fingers are divided into bone segments – the first four fingers of each hand having 12 segments and the thumb having two, lead to 14x2=28 total *phalanges* which is the technical term for each segment. Likewise the feet have 28 corresponding phalanges, 14 in each foot. We encounter the perfect number 28 several times in the course of this book, as it is significant in several contexts – one prominent link being the 28 day cycle of the moon. Chapter 8 revisits the number 28 within the framework of 'mystical numbers' and covers some of its curious properties.

Body postures, whether in continual motion in the form of dance, or in the realm of yoga, have been studied for centuries prior to our modern times. The ancient tradition of the 'whirling of the dervishes' also has a long and glorious history. Since the death of the great poet Jalaluddin Rumi (1207-1273) who founded the Mevlana Sufi order, the tradition of the *sema* ceremony and whirling sufi dance has endured in his hometown of Konya in Turkey, as a commemoration. The spiral motion in the graceful trance-inducing whirling of the Sufi dance and also in the many yogic asanas or postures practiced by Yogis, suggest the presence of the golden ratio. It is as though the body in its most harmonious, balanced state, be it static or in motion, also aspires to the harmonic pattern that is widely prevalent throughout the universe.

Patanjali, the Indian sage believed to be born between the second and fourth centuries C.E., has been attributed as the author of the *Yoga Sutras*, a classical yoga text on theory and practice, his most famous work containing 196 aphorisms. For nearly 700 years this seminal text fell into obscurity and was only revived in the 19th century by Swami Vivekananda, among others. In recent times, yoga and particularly the physical '*asana*' aspects of it in the form of Hatha Yoga and Ashtanga Yoga have become immensely popular throughout the world. Many of the yogic *asanas* not only claim to be beneficial to the general health and well-being of the practitioner, but align with the spiral pattern seen in the Fibonacci sequence and give rise to the golden ratio. It is believed that the sacred geometry and the golden ratio embedded in the many postures and movements of the human form demonstrate a sense of beauty and also serve to harmonise the body through the geometry of natural balance.

Leonardo Da Vinci (1452-1519) was an Italian polymath renowned for his exceptional talents as a painter, mathematician, engineer, architect, and sculptor, aside from being involved in several other areas like cartography. Da Vinci's most famous drawing is known as the *Vitruvian Man* (c. 1490) which is a detailed study of the proportions of the human body and also serves as a seamless link between the

Arts and the Sciences through a single piece of work. His extraordinary portrayal of the human form must have arisen from his detailed knowledge of the inner mechanisms of the body – the subcutaneous workings of the tendons, muscles and sinews and the network of structures that comprise the human anatomy. Leonardo apprenticed as a student under Andrea del Verrocchio and learned about the detailed topography of the human body and the various parts of the skeletal and muscular structure. As previously noted, it is over 500 years since Leonardo da Vinci performed the first scientific examination of the human body in Florence. His curiosity was initially fueled by his desire to paint the human form more accurately; a feat he thought he could best achieve through dissection. His artistic sketches were of such high calibre, not only in the accuracy of the drawings but also in their clarity of perception that they led to the introduction of cross-sectional diagrams in anatomy.

The extraordinary geometry of natural form, particularly that of our own species, must fascinate children when they first discover their own skeletal and muscular structure. It is indeed a thrilling moment when a child discovers his or her rib cage and tries to count how many emergent rib bones there are. The spinal column is also sure to intrigue any amateur student of anatomy or physiology even before they learn these categories of biology. Prudence would serve the children well, however, to not get too curious about the workings of the human body. They might inadvertently injure themselves while experimenting with a sharp thorn merely to observe a drop of blood. Nearly everyone remembers the sight of blood when their shin was first scraped. It was indeed a matter of amazement to notice how a wound quickly healed even without intervention.

It is this curiosity of human form and anatomy that gives rise to a mathematical assessment of proportion. Da Vinci did show knowledge of Vitruvius, despite not adhering to the exact description in the ancient text reproduced below from de Architectura:

For the human body is so designed by nature that the face, from the chin to the top of the forehead and the lowest roots of the hair, is a tenth part of the whole height; the open hand from the wrist to the tip of the middle finger is just the same; the head from the chin to the crown is an eighth, and with the neck and shoulder from the top of the breast to the lowest roots of the hair is a sixth; from the middle of the breast to the summit of the crown is a fourth. If we take the height of the face itself, the distance from the bottom of the chin to the under side of the nostrils is one third of it; the nose from the under side of the nostrils to a line between the eyebrows is the same; from there to the lowest roots of the hair is also a third, comprising the forehead. The length of the foot is one sixth of the height of the body; of the forearm, one fourth; and the breadth of the breast is also one fourth.... For if a man be placed flat on his back, with his hands and feet extended, and a pair of compasses centred at his navel, the fingers and toes of his two hands and feet will touch the circumference of a circle described therefrom. And just as the human body yields a circular outline, so too a square figure may be found from it. For if we measure the distance from the soles of the feet to the top of the head, and then apply that measure to the outstretched arms, the breadth will be found to be the same as the height, as in the case of plane surfaces which are perfectly square.

Although Da Vinci indicates direct knowledge of Vitruvius, his own drawing of 'The Vitruvian Man' does not stick to the description in the original text. He notices that the square cannot have the same centre as the circle and therefore the centre of the human body must be located at the groin. This is a unique departure from previous illustrations and he also diverts from Vitruvius by drawing the arms raised to a position such that the fingertips are leveled with the top of the head.

Leonardo Da Vinci's drawing is reproduced here in the figure below, and indicates sixteen different poses arising out of the combination of arm and leg positions. The pose with the arms drawn out straight and the feet joined together is inscribed in the superimposed square, whereas the pose with the arms and legs spread out is inscribed in the superimposed circle.

Over time, many scientists and mathematicians have pointed out the repeated occurrence of the golden ratio in human proportions, approximated by the numerical value 1.6180, which we shall revisit in later chapters. In a recent study (2012), Takashi Ida of the Nagoya Institute of Technology, Japan, concludes that the ratio of the radius of the circle to the side length of the square in Da Vinci's 'Vitruvian Man' was intended to be 137 / 225 = 0.6088..., and not the golden ratio.

Although the drawing of the Vitruvian Man at first glance seems to merely indicate the impressive human proportions of bilateral symmetry, it actually suggests the solution to the

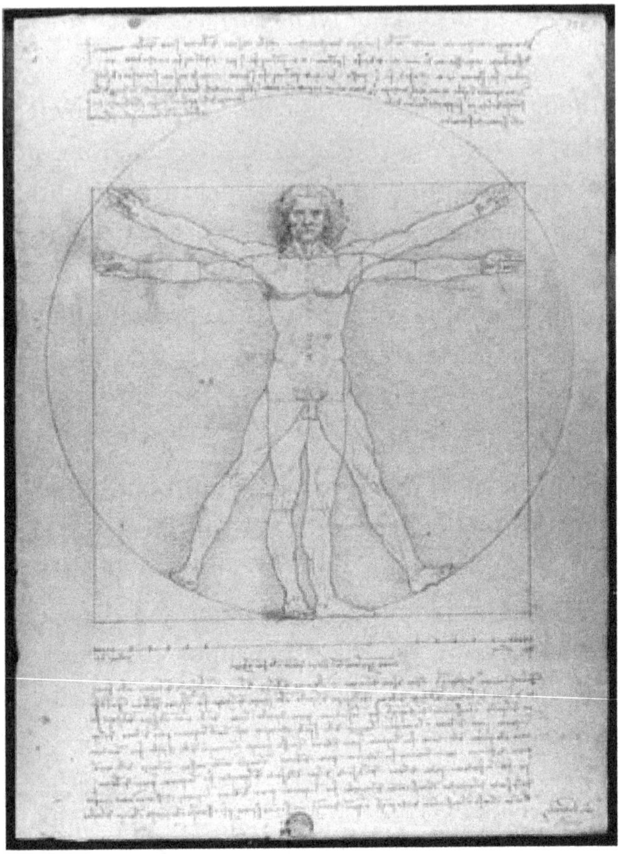

Da Vinci's Vitruvian Man

problem of the 'squaring of the circle' – an ancient mathematical challenge that remained without a satisfactory solution until much more recently. Even while playing with sticks one may notice that creating a radial disk out of them can be dismantled to produce a square shape by stacking the sticks linearly instead of radially. Thus, one may notice that better and better approximations are arrived at with a larger number of sticks or with thinner sticks. These experiments are forerunners to the idea of iterations giving rise to greater accuracy when solving the problem – or in other words these are the initial building blocks of calculus itself.

The mathematical fractal like properties of the inner workings of the human body include the circulatory system and the branching of the lungs and kidneys. One may expect a self-similar fractal-like branching distribution system within the human body, similar to the root systems in trees which show branching behaviour in order to spread nutrients efficiently through the entire system. Specifically, researchers tend to classify the respiratory system as exhibiting fractal behaviour because of its geometrical and material properties. An important study by Clara Ionescu et al (2009) extended previous work on human and animal lung parenchyma assuming a power-law behaviour, common to all fractals, of the visco-elastic properties in the soft biological tissue. The paper attempted to verify if the 'fractional order operator' was also related to the structure of the respiratory tree. Ultimately, it was shown that the dichotomous and recursive, fractal-like properties of the respiratory system lead naturally to the fractional-order operators (or the particular power law behaviour).

The presence of structure in the small scale being repeatedly seen on a larger scale is displayed in many natural phenomena, like the branching of lightning or the spreading of mycorrhizae fungi beneath the soil. It is therefore not surprising that this 'networking' geometric structure is found in the human body as well – from the beauty of the iris to the capillary distribution of the circulatory system. Overall, the human form appears to have been cast in a way that is both mathematically elegant in its interior fractal-like structure, and proportionately full of symmetry and grace in its exterior manifestation. One is reminded of the line from William Blake's "The Tyger," 'What immortal hand or eye, Could frame thy fearful symmetry.' In the case of the Vitruvian man, it is obvious that Da Vinci has attempted to neatly 'frame' the symmetry of man, of the human form in a cleverly iterative way. Interpreting Blake in his daring verse, one senses that it is virtually impossible to perform a complete framing to capture the incredible majesty of the tiger. Not even the Immortal Mind, let alone the mortal one could frame it. Consequently, are we left with the impossibility of our task of framing the wondrous symmetry of Human Form, no matter how many iterations or mathematical simulations are employed in its construction?

CHAPTER SIX

FIBONACCI SERIES IN NATURE

"To see a world in a grain of sand and heaven in a wild flower. Hold infinity in the palm of your hand and eternity in an hour."

—William Blake

The untamed beauty of wildflowers had surely touched us when we were strolling along a country path or a wild garden. We may recall the forgivable urge to pluck a flower or two and inspect them carefully, counting their petals and witnessing their magical arrangements. The number of petals in most flowers tend to follow a number pattern which we call the Fibonacci sequence of numbers, where each digit arises out of adding the previous two, starting with 1. Thus, the sequence of the following numbers, 1,1,2,3,5,8,13,21,34,55,89,144... was constructed by adding the previous two numbers. When children grow up with poetic traditions of song, rhythm and rhyme, they are seemingly unaware of the signature of the Fibonacci sequence in their carefree cultural activities. It seems a shame that this sequence is often not included in school curricula, and students only ever accidentally learn of it, if at all.

When young learners are left to themselves, one ought to keep open the possibility that many discoveries might naturally occur from mere inspection. Introducing children to Fibonacci numbers at the earliest opportunity would also alert them to their propensity in nature across both natural and artificial structures. One easy and fascinating example is of the hexagonal bracts on a pineapple which form three separate-direction spirals. For instance, they may well be 5, 8, and 13 spirals, or even higher order ones that also happen to be Fibonacci numbers. In another example, if one inspects pine cones, one will find that they have two distinct direction spirals. The number of spirals in each direction tend to be two successive Fibonacci numbers. Daisies are common flowers that also exhibit Fibonacci numbers in the numbers of petals on display. A whole array of flowers like marigolds, asters, roses, buttercups and lilies are famously known to exhibit Fibonacci numbers in their petal counts.

The sunflower, similar to pine cones, has a variety of different spiral numbers, the older the sunflower the greater the Fibonacci numbers in the spiral count. Typically, the spiral directions are opposite – clockwise and anti-clockwise – and count two consecutive Fibonacci numbers. These spiral packing arrangements seem to be the most efficient in nature, it turns out, with all the seeds of the same size being packed together without too much crowding near the centre or sparseness near the edges.

Generating the Fibonacci numbers using our initial rule of adding the previous two numbers to get the current one, starting with 0 and 1, quickly give us,

0 + 1= 1

1 + 1= 2

1 + 2= 3

2 + 3= 5

3 + 5= 8

5 + 8= 13

8 + 13= 21

13 + 21 = 34

21 + 34 = 55

34 + 55 = 89

55 + 89 = 144 ... which are the first twelve of an infinite sequence. As it turns out, adjacent Fibonacci numbers are relatively prime, which means they have no common divisors. An interesting point to note is the absence of cubes and squares in the sequence; 8 is the only non-trivial cube (2x2x2) and 144 is the only non-trivial square (12x12).

Another curious feature of the Fibonacci sequence is that the sum of ten consecutive numbers of the sequence happens to be divisible by 11; this property holds true throughout the series. We might randomly check say numbers F_{10} through to F_{19} giving us the result indicated below:

$$F_{10} + F_{11} + F_{12} + \ldots + F_{19} = 55 + 89 + 144 + 233 + 377 + 610 + 987 + 1597 + 2584 + 4181 = 10{,}857$$

which is evenly divisible by 11 to give 987, another Fibonacci number.

Some of the properties of the Fibonacci sequence are indeed profound. Addition of consecutive Fibonacci numbers present the subsequent one; but what happens if we square each of the Fibonacci numbers? The following sequence emerges: 1,1, 4, 9, 25, 64... Adding these numbers by picking any two consecutive ones give us the original sequence shifted over by one place; in other words, 2,3,5,8,13 and so on. Surprising cyclical patterns arise if you add the squares of the Fibonacci numbers in batches of ones, twos, and threes. We obtain 1+1 =2, 1+1+4=6, 1+1+4+9=15, 1+1+4+9+25=40...and so on. These numbers have the Fibonacci sequence embedded in them since 2,4,6,15,40... are multiples of the Fibonacci numbers.

2=**1x2**

4=**2x2**,

6=**2x3**,

15=**3x5**,

40=**5x8**...each of these highlighted digits on the right are themselves Fibonacci numbers.

Furthermore, if we express the sum of the squares making up the smaller rectangles, we get the pattern produced below. Notice again the presence of the Fibonacci numbers on the right hand side.

$$1^2+1^2=1\text{x}2$$

$$1^2 + 1^2 + 2^2 = 2\text{x}3$$

$$1^2 + 1^2 + 2^2 + 3^2 = 3\text{x}5$$

$$1^2 + 1^2 + 2^2 + 3^2 + 5^2 = 5\text{x}8$$

$$1^2 + 1^2 + 2^2 + 3^2 + 5^2 + 8^2 = 8\text{x}13$$

$$1^2 + 1^2 + 2^2 + 3^2 + 5^2 + 8^2 + 13^2= 13\text{x}21 \ldots$$

Early familiarity with the Fibonacci sequence was expressed by ancient Indian mathematicians in the Sanskrit poetic tradition and in the context of classical Indian music. The poet Pingala (c.450-200 B.C.E.) elaborated rhythmic structures in his verses where patterns of long (L) syllables of two units duration were juxtaposed with short (S) syllables of one unit duration. The contemporary professor of Mathematics and Field's medalist (2014) Manjul Bhargava mentions the importance of patterns in mathematics in the Field's Symposium Lecture given in 2016. He emphasises the presence of patterns that arise naturally through poetry and music by exploring the example of short and long beats, of one or two units in length, via Sanskrit poetic prosody. Bhargava relates the earliest findings of Virahanka (c. 770 B.C.E), the poet and linguist who gave early indications of mastery over the poetic form and noted how poetic beats and their counting leads to what is now recognised as the Fibonacci sequence. Virahanka built his work on prosody on the previous works of Pingala from the 4th century B.C.E. Much later, in the 12th century, the work of Gopala was built on these early treatises.

In the following passage, Gopala (c.1135) speaks of,

Variations of two earlier metres (is the variation)...For example, for (a metre of length) four, variations of metres of two (and) three being mixed, five happens. (Examples 8, 13, 21 are worked out)...In this way, the process should be followed in all matra-vrttas (prosodic combinations).

Bhargava attributes his interest in Sanskrit poetry and the tabla instrument to his early years in India which he spent with his grandfather, a Sanskrit scholar, poring over ancient Sanskrit texts. His exposure to ancient texts, Sanskrit poetry in particular, ultimately led to his forming fascinating and unexpected links between seemingly unrelated fields. Bhargava also mentions the poet, philosopher and grammarian Hemachandra who described the Fibonacci sequence in 1150, as emerging from cadences of length n, formed by adding a short sequence of cadence $n-1$ to a long sequence of length $n-2$. Sanskrit poetry consists of syllables of long and short rhythms. A long rhythm lasts twice as long as a short rhythm, so if the short rhythm is one unit or one beat, the long is two beats. As an illustrative example, we are asked to consider how many rhythms can one construct out of eight beats, consisting of long and short syllables. The possible outcomes can be labeled thus,

long long long long

short short short short short short short short

short long long short long

long long short short long...

The poets went on to state the question more mathematically. In how many ways can one write a number n as a sum of ones and twos? The response was simply noticing the pattern that developed from the simplest instance of $n=1$, therefore if there is only one beat allowed how many ways can the rhythm form?

$n=1$	1 way: a short beat (there is no room for long)
$n=2$	2 ways: long and short short
$n=3$	3 ways: long and short, short short short, short and long
$n=4$	5 ways: long short short, long long, short short long, short long short and short short short short

n =5 8 ways: short short short short short, short short short long, short short long short, short long short short, long short short short, long long short, short long long, long short long ..., and the pattern continues. One can replace short and long with the number 1 and 2 to give further clarity. So, in how many ways can one form the number *n* using ones and twos? Then,

n=1 1 which means just one way

n=2 1+1 = 2 ways

n=3 2 +1, 1+1+1, 1+2 = 3 ways

n=4 2+1+1, 2+2, 1+1+2, 1+2+1 and 1+1+1+1 = 5 ways

n=5 1+1+1+1+1, 1+1+1+2, 1+1+2+1, 1+2+1+1, 2+1+1+1, 2+2+1, 1+2+2, 2+1+2

amounts to 8 ways. This pattern may continue indefinitely. However, we notice the beginnings of the familiar Fibonacci sequence, 1,2,3,5,8. This series can now also be called the Virahanka-Fibonacci sequence or the Hemachandra numbers, re-named to honour some of the many grammarians, poets and mathematicians who had discovered these patterns centuries before they became more popular in Europe.

Outside of India the Fibonacci sequence was first noted in *Liber Abaci* where Leonardo of Pisa or Fibonacci (c.1170) considers the growth of an idealised rabbit population where each breeding pair mates at the age of one month. At the end of the second month the second pair produce another pair of rabbits. The rabbits never die and continue to breed indefinitely. Considering this scenario Fibonacci poses the question: how many pairs will there be in a year's time? At the end of the first month there is still only one pair, but by the second month there are two pairs in the field. By the third month the original pair produce a second pair but the second pair only mate without breeding so the total number of pairs in all are 3. At the end of the fourth month the original pair has produced a new pair again and the pair born two months prior also produces their first pair making 5 pairs altogether. At the end of the 'n^{th}' month the number of rabbit pairs is equal to the number of mature pairs in the $(n\text{-}2)^{\text{th}}$ month plus the number of pairs in the previous $(n\text{-}1)^{th}$ month. Therefore, the number in the nth month is the nth Fibonacci number. Written succinctly,

$$R_n = R_{n-2} + R_{n-1}$$

Many of the properties of the Fibonacci sequence have lead to widespread applications in areas of mathematics and other related subjects – from optimisation problems arising in economics to Fibonacci coding in computer science and from parallel computing to the technical analysis of long term behaviour in the stock market. More recent applications include modern methods in pattern behaviour predictions called Fibonacci retracements in cryptocurrency technical analysis and trading. The presence of the sequence is so ubiquitous that we even find it in classical music, both Eastern and Western.

In Western music, an octave on the piano consists of 13 notes; eight are white keys and five are black keys. A scale is composed of eight notes, of which the third and fifth notes create the foundation of a basic chord. In a scale, the dominant note is the fifth note, which is also the eighth note of all 13 notes that make up the octave. We can see the presence of the Fibonacci sequence from these numbers: 5, 8 and 13, and also a crude golden ratio approximation when we divide 13 by 8. In a recent article in the Journal of Applied Acoustics, authors Sinith, Tripathy and Murthy (October, 2020) suggest that the Fibonacci series and music are related in many ways. Even though Indian Classical Music is known to be based on Just Intonation (tuning of musical instruments as whole number ratios) rather than equal tempered, not much attention has been given to inflections on the notes, which are an integral part of the notes themselves. 'In this work an interesting relation between Fibonacci series and Just Intonation in classical *ragas* has been discovered. Further, discrete versions of pitch contours based on a table derived using the Fibonacci series are introduced for these *ragas*. These pitch contours can effectively represent the melody and can be a tool for music information retrieval. A pitch distribution based on the Fibonacci series is also introduced in this research study. Raga recognition is performed using this distribution and a recognition rate of 95.3 percent is obtained for the typical 17 *ragas* considered for the experiment.' The recognition rate is higher compared to reported work in raga recognition.

One of the most important relations that arise out of the Virahanka-Fibonacci series is the ratio between two consecutive numbers in the sequence. A surprising and yet powerful relation emerges called the Golden Ratio or the Golden Mean which has been determined to be:

$$\text{phi} = \varphi = 1.6180339887498948482045868343656\ldots$$

When Fibonacci numbers are taken as quotients in consecutive pairs, they approach the golden ratio with greater accuracy the further along in the sequence. Early in the sequence if two consecutive numbers, say 21 and 34 are taken, their ratio (quotient of the greater number to the smaller) would equal 34/21=1.619047619047619. However, if a consecutive pair later in the series, for instance, 377 and 610 are taken and the bigger number is divided by the smaller, we get closer to the true number as indicated by the following relationship,

$$610/377=1.618037135278515.$$

Likewise, the 22nd and 23rd Fibonacci numbers are 17711 and 28657 respectively. Thus, the following ratio is remarkably close to the actual golden ratio,

$$28657/17711= 1.618033990175597.$$

The subject of much of the next chapter will be the golden ratio itself and how it seems to be omnipresent across fields: art, architecture, poetry and music to the behaviour of many natural and observational phenomena. Aside from the apparently boundless presence of Fibonacci numbers in flower petals and seed spirals, in the arrangement of leaves arising from the dynamics they seem to display and in a variety of fruits and plants, animal spots and sea shell spirals, these numbers and the golden ratio are also present in countless works of human endeavour. Numerous temples, mosques, churches and religious places of worship including prominent monuments and buildings have harnessed the natural proportions present in the golden ratio. The presence of this ratio in so many areas have lead several observers, mathematicians, geometers and architects, to call the study of this subject 'Sacred Geometry.'

CHAPTER SEVEN

SACRED GEOMETRY

"Let those who understand geometry enter here."

—Sign over the door of Plato's Academy

Plato founded his academy in Athens in 377 B.C.E aimed for the study of mathematics, political theory and philosophy. Salient among his other written works is the Timaeus, written towards the end of his life, in which he covers a variety of topics that include the geometric progression of harmonics and also the study of a subset of polyhedra later called the 'Platonic Solids.'

A platonic solid is a three-dimensional shape where each face is the same regular polygon, and the same number of polygons meet at each vertex or corner. The faces of each platonic solid must be identical and the angles at the vertex must be less than 360 degrees. Therefore, there are only five platonic solids – namely, the tetrahedron where three triangles meet at each vertex, the octahedron, where four triangles meet at each vertex, the icosahedron where five triangles meet at each vertex, the cube, where three squares meet at each vertex and the dodecahedron where three pentagons meet at each vertex. These solids are essential to Plato's philosophy in which each of the four classical elements, earth, air, water and fire, were associated with the cube, the octahedron, the icosohedron and the tetrahedron. These polyhedra have been extensively used in architecture and art, particularly in relation to sacred and religious buildings and paintings. Other special ratios and patterns also feature in the construction of these sacred monuments and the mathematics used in their construction is called 'Sacred Geometry.'

The golden ratio, the Virahanka-Fibonacci sequence and other forms of related symmetric designs appear not only in nature, but in music, art, photography and architecture; in edicts, temples, places of worship and dwelling quarters around the world. The ubiquitous presence of many of these mathematical symbols and their representations throughout history, in various places, from ancient pyramids to Japanese temples, have lead admirers of such art and architecture to call the mathematics behind these representations 'Sacred Geometry' which ascribes symbolic meaning to particular geometric shapes and designs, proportions and forms. This emphasis on the 'sacred' also seems to suggest that God is the great geometer of the universe.

The golden ratio, and other proportions arising from the Virahanka-Fibonacci sequence, have been liberally used in the building of ancient monuments, tombs and religious places of worship. Along with the golden ratio, another symbolic pattern that appears nearly everywhere, regardless of culture, is called the flower of life shown in the diagram below.

Flower of Life

The flower of life pattern can be extended outwards in concentric hexagonal intersecting circles placed with six fold symmetry. Historically, this pattern appears in several representations through the ages such as in Phoenician art forms from the 9th century B.C.E. Among the earlier representations is a mythological scene of a sphinx frieze etched in a cup from the Cypro-Archaic period (c. 7th - 8th centuries B.C.E). Floor decorations, dated to around 645 B.C.E., from the northern Iraqi palace of King Ashurbanipal, which are now housed in the Museum of Louvre in Paris, also depict this timeless 'flower of life' design. Nineteen-circle diagrams drawn in red ochre have been discovered in the temple of Osiris at Abydos in Egypt and a similar geometrical pattern on the surface of a ball held by the male imperial guardian lion is located at the ancient Gate of Supreme Harmony in the Forbidden city of Beijing in China.

The importance of the flower of life has also been reiterated in the drawings of Leonardo Da Vinci, who extensively studied the form of this design and the many mathematical properties that are embedded within its derivatives. The Platonic solids including spheres, a torus and the golden ratio phi are the different geometrical features that can be extracted from the underlying mathematics of the flower of life pattern.

Beside the flower of life, the irrational numbers $\sqrt{2}$ and $\sqrt{3}$, along with the golden ratio are frequently found in the geometry of architectural structures throughout history. The $\sqrt{2}$ or 1.414213562...appears in a right angled triangle with side lengths of one unit and perfectly embodies the principle of the 'whole being more than the sum of the parts;' in other words, the identical vertical and horizontal sides giving rise to the hypotenuse which is greater than either. As we discovered in earlier chapters, patterns, structural symmetry and self-replication are found in nature starting from the tiniest forms to the largest geometrical archetypes like the spiral galaxies. Regular geometrical forms like the circle with its relationship with pi (3.14159261...), and its higher order three-dimensional version, the sphere, are repeatedly represented in the realm of sacred geometry. Along with these regular shapes, the square is also commonly represented together with the three-dimensional cube which can house the platonic solids – the sphere, the pyramid, the dodecahedron and the icosohedron.

As we have witnessed in the previous chapter, the golden ratio, 1.618033...which is also an irrational number and is borne out of the limit of the ratios of two consecutive numbers of the Virahanka-Fibonacci sequence, has some unique properties and embodies interesting features that explain its widespread presence in architecture, art, photography and music. Some of its mathematical idiosyncrasies can be summarised thus, for the sake of collating these properties together:

- $\varphi = 1/\varphi + 1$
- φ^2-φ-1=0; this means $\varphi=(1+\sqrt{5})/2$ and $(1-\sqrt{5}/2)$ or 1.618033 or -0. 618033.
- $\varphi = \varphi^2 - 1$; therefore $1 + \varphi = \varphi^2$; $\varphi + \varphi^2 = \varphi^3$; $\varphi^2 + \varphi^3 = \varphi^4$; *ad infinitum.*
- $\varphi = (1 + \sqrt{5}) / 2$ from quadratic formula, $1 + \varphi = \varphi^2$
- $\varphi = 1 + 1/(1 + 1/(1 + 1/(1 + 1/(1 + 1/(1 + 1/...)))))...$
- φ = (sec 72)/2 =(csc 18)/2 = 1/(2 cos 72) = 1/(2 sin 18) = 2 sin 54 = 2 cos 36 = 2/(csc 54) = 2/ (sec 36)
- ratio of segments in 5-pointed star (pentagram) considered sacred to Plato and Pythagoras in their mystery schools. Each larger (or smaller) section is related by the phi ratio, so that a power series of the golden ratio raised to successively higher (or lower) powers is automatically generated: φ, φ^2, φ^3, φ^4, φ^5, etc.
- φ = apothem to bisected base ratio in the Great Pyramid of Giza.

A special feature of $\sqrt{3}$ is its presence in what is known as 'the Vesica Piscis' – an area formed by the intersection of two circles or spheres whose centres touch. The shape of the human eye itself can be considered to be a Vesica Piscis. The spiritual significance of "seeing eye to eye" to the "mirror of the soul" was highly regarded by numerous Renaissance artists who used this form extensively in art and architecture. The ratio of the axes of the form is the square root of 3, which alludes to the deepest nature of the triune which cannot be adequately expressed by rational language alone.

Sacred Geometry in art and architecture employs the spiral generated by a recursive nest of Golden Triangles (triangles with relative side lengths of 1, φ and φ). It is also the classic shape of the chambered nautilus shell. The creature building this shell uses the same proportions for each expanded chamber that is added; growth follows a law which is everywhere the same.

The vast arena of fractal and recursive geometries also covers many elements of sacred geometry. As highlighted previously, repeated patterns are found in ancient monuments, mausoleums, religious places of worship and grand palaces, particularly in the world of Islamic architecture. The concept of the individual containing the features of the whole and vice versa is indeed a profound truism that applies across many fields where these features crop up. Certainly crop circles famously follow fractal designs and their study has been receiving much attention lately, both in popular literature and in mathematical theory.

Interesting ratios of whole numbers also repeatedly appear in edicts and monuments, among other buildings. The 3/4/5, 5/12/13 and 7/24/25 triangles are examples of right triangles whose sides are whole numbers. The 3/4/5 triangle is contained within the so-called "King's Chamber" of the Great Pyramid, along with the $2/3/\sqrt{5}$ and $5/\sqrt{5}/2\sqrt{5}$ triangles, utilizing the various diagonals and sides.

Pyramids at Giza, Egypt.

The five Platonic solids (tetrahedron, cube or (hexahedron), octahedron, dodecahedron and icosahedron) are ideal, primal models of crystal patterns that occur throughout the world of minerals in countless variations. These are the only five regular polyhedra, that is, the only five solids made from the same equilateral, equiangular polygons. To the Greeks, these solids symbolized fire, earth, air, spirit (or ether) and water respectively. The cube and octahedron are duals, meaning that one can be created by connecting the midpoints of the faces of the other. The icosahedron and dodecahedron are also duals of each other, and three mutually perpendicular, mutually bisecting golden rectangles can be drawn connecting their vertices and midpoints, respectively. The tetrahedron is a dual to itself.

The Prince's Foundation School of Traditional Arts, U.K., is an institute that keeps alive the study of Sacred Geometry and the traditional arts and architecture through its many programmes and courses including a master's study leading to a degree in the subject and recently the MPhil and PhD degrees. It is here, that several courses on Islamic architecture, sacred drawings and traditional art and print-work are offered in the hope that these traditional arts do not get swept away into oblivion, but rather endure as living and thriving practice even in modern times. The artist and professor of architecture, Keith Critchlow (1933-2020), famous for his extraordinary lectures on Chartres Cathedral, founded the school. From experimenting in the construction of geodesic domes using lightweight materials like palm, bamboo and aluminium to re-designing the minbar of Saladin in the Al-Aqsa mosque in Jerusalem, Critchlow has used his extensive knowledge of polyhedral shapes and geometrical patterns in re-instating a lost tradition of sacred building. His seminal and pioneering study of the geometry underpinning Islamic art has been crafted into the book 'Islamic Patterns: An analytical and cosmological approach,' (1983). I have myself had the privilege of living in the Krishnamurti Centre in Hampshire, which was designed by Critchlow using massive wooden beams, red bricks, grey flint stones gathered locally and clay tiles, all without the use of nails. I felt that this building embodied the coming together of arithmetic and geometry, embracing the harmony of the music of the spheres in its graceful and yet unobtrusive presence.

In South India, temples have preserved the old traditional symmetric design features that carry deep symbolic meaning and follow strict mathematical guidelines. The temple

Front portion of Saladin's minbar

building styles of the south possessed several distinctive features and have been mentioned as one of the three styles of temple building in the ancient book *Vaastu Shastra*. Many of the temples remain as heritage sites and places of religious pilgrimage and worship to this day in the southern states of Tamil Nadu, Karnataka, Kerala, Andhra Pradesh and Telengana. The appearance of self-similar patterns and fractal-like self-similarity features abound in the walls and interior and exterior structures of the Mahadeva temple in Khajuraho, Central India. South Indian temple architecture typically is an assemblage of porches and towers that terminate in a shikhara or spire. The uniquely tactile quality of the stone and also the character of the symmetric designs in the high relief carvings of the figurines engraved throughout the Mahadeva temple display not only the decorative quality and ornamentation of style but also describe the syntheses of fractal features of the temple from Hindu cosmology and philosophy. A study by Iasef Md Rian et al. (2007) carefully outlines the mathematical and cosmological significance of the fractal features of the temple.

The most distinctive part of the temple are the lofty curved towers above the sanctuary known as the *shikhara* which is crowned by an *amalaka*, a ribbed circular motif, and a pot finial, known as the *kalasha* – a sacred pot containing the precious elixir *amrita*. The amalgamation of the form of lotus blossom and that of the mountain has been frozen into the form of multiple recursive archetypes in the *shikhara* itself. Replicas of the towers form the prominent characteristics of the temple.

The fractal like floral pattern of the panels in the ceiling of the Mahadeva temple depict the cosmos in such a way that the flowers that are shown appear to blossom from buds recursively.

The floral pattern of panels in the temple ceiling is most noteworthy for depicting the cosmos, where again mathematical fractal self-similarity is used for illustration. Fractal patterns of the plan of the temple, replicas of detailed towers, and the rhythmic growth of horizontal friezes, create a harmony of fractal characters in the entire complex. Again, fractal geometry symbolises the synthesis of Hindu cosmology in a language that translates the beauty of complexity of creation and destruction in endless repetitive cycles, where Brahma is the creator, Vishnu the sustainer and Shiva the destroyer.

Kandariya Mahadeva temple detail, Khajuraho, Madhya Pradesh

Along with the temple as the *sanctum sanctorum* for supplication, worship and spiritual practice, other forms of worship in the ancient Indian tradition include the drawing of kolam patterns on the ground – traditionally using powdered rice. This tradition has been a practice in south India presumably for thousands of years, going back to the Indus Valley Civilization, as mentioned in a paper by Rethinasamy Kittappa (2019). Features of these fascinating designs and patterns include knots, curves and lines, geometrically placed with artistic regularity to create a spiritual offering as well as a mathematical artifice. The underlying relationship between Hindu cosmology and fractal theory is manifested in temples where fractal geometry acts as the *language.* Many *kolam* designs and sacred motifs include the images of animals and birds to symbolise the unity of man and animal. Cosmological variants and representations of the sun, moon and zodiac signs are also used, along with geometrical shapes like squares, circles and triangles – the downward pointing triangle representing woman and an upward pointing triangle representing man. This art of *kolam* creation has been passed on for hundreds of generations and continues unabated in several traditional families in South India to this day. *Kolams* are typically set in grid patterns, with open strokes running along each dot. Sometimes the strokes are connected between the dots variously representing flowers or animals. Often patterns are arranged in a radial fashion called a 'lotus' and maybe dyed in bright colours. Vijaya Nagarajan, in her book *Feeding a Thousand Souls: Women, Ritual and Ecology in India, an Exploration of the Kolam* (2018) considers the ancient Hindu mythological idea of individuals having a 'karmic obligation' to 'feed a thousand souls.' The rice flour *kolam* designs are offerings not only of prayer, but food – for the many ants, insects, birds, and other creatures that happily live among us. In this unusual expression of mathematical ideas, a dual offering to divinity and nature is presented as a ritualistic start of the day. Furthermore, according to Nagarajan, the *kolam* is among the uniquely embedded indigenous traditions that contributed to the flowering of western mathematical traditions.

Buddhist *mandalas* are similar to *kolams* in their patterned geometrical configuration of symbols. In the different Buddhist traditions, *mandalas* are utilised as spiritual aids in meditation and in focusing the attention of the practitioner. In one type of *mandala*, the visualisation of Vajrayana traditions of Buddhism is conducted through the weaving of silk tapestry with gilded paper depicting elements like crowns and jewels via the *Thangkha*. A mandala can also represent the entire universe through the depiction of Mount Meru, as the *axis mundi* in the centre. In the ancient tradition of the 'sand *mandala*,' as a meditation upon the impermanence of life, after days or weeks of painstaking artistry and labour in creating intricate patterns of sand, the mandala is destroyed by brushing the sand together in a pile. It is then spilled into a body of running water to spread the blessings.

Thangkha artist painting an intricate design

Buddhist architecture also often applies the *mandala* as the blueprint for temple complexes and *stupas* which are sacred mounds of relics. Notably, the *mandala* architecture in Borobodur, Java, built in the 9th century forms a large *stupa* surrounded by self-similar replications arranged on terraces forming a stepped pyramid. When viewed from above, the pattern takes the form of a *mandala* representing Buddhist cosmology and the nature of mind. The essence of Buddhist philosophy is also captured in the *pagodas* and temples devoted to the worship of the Buddha. The Tibetan style of symmetric designs and patterns are etched into the weaving of *thangkhas* – hangings and tapestries that display intricate sacred patterns, designs and human and animal forms, and depict cyclical stories that serve as teaching tools. The worship and practice of the religion is intertwined with the weaving process and therefore the very act of weaving cannot be separated from the theology and practice of the Buddhist religion and teachings or indeed from the geometrical and mathematical significance of the practice.

In Japan, a period of history known as the 'Edo' arose in the early seventeenth century under singular circumstances that lead to the flourishing of 'Japanese temple mathematics and geometry.' This unique time marked the creation of artistic wooden tablets adorned with beautiful geometric problems that were simultaneously works of art, religious offerings, and a record of what might be loosely termed as 'folk mathematics.' For historical reasons, the isolation from the Western world was nearly complete although there was limited trade with Korea and China. The political isolationist policies were apparently so successful during this period of

Edo history that the elimination of foreign conflicts during this time frame came to be known as 'The Great Peace.' The creators of these *sangaku*, over two centuries – Japanese mathematicians, professionals, amateurs, women and children – invented what was the equivalent of 'mathematical' stained glass windows in churches. The collective creation of these artefact and their sacred manifestation in the temples themselves could be labelled as *temple geometry sacred mathematics*.

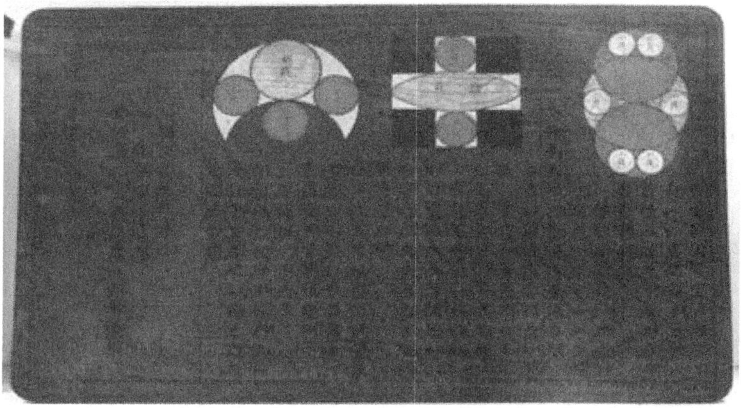

Sangaku geometrical problem box dedicated to Konnoh Kachimangu (Shibuya, Tokyo, 1859).

The *sangaku* were created by a profoundly different aesthetic than the Greek inspired designs found in Western geometry books. In a remarkable way, the traditional Japanese geometers who created Sangaku saw their mathematical world through different eyes and sometimes solved mathematical problems in distinctly non-Western styles. To learn Japanese mathematics is to learn another way of thinking. The *sangaku* were artistic and geometric problems, solved and offered to the Japanese temple as a mark of respect; as a gift. Most contained only the final answer to the mathematical or geometric problem. The presenters of the *sangaku* obviously took the spiritual and aesthetic aspects of the posed problem seriously. An exhaustive and impressive treatise called Sacred Mathematics, about Japanese temple geometry, by Fukagawa Hidetoshi and Tony Rothman (2008), gives a window of insight into the world of mathematical thinking that arose from a spiritual foundation within the broader cultural and historical setting of that period.

The early history of Japan is inextricably bound up with that of China, from which much of its culture of mathematics was imported. The use of the Chinese abacus or *soroban* is evident from an old Chinese book – Cheng Dawei's famous Suanfa Tong Zong or the systematic treatise on Arithmetic (China, 1593). This book made its way to Japan thenceforth and influenced much of Japanese sacred geometry and artistic puzzle creation. The concept of large and small numbers and numerical problems such as the calculation of pi also appeared in the book 'Sanso' or 'Stack of Mathematics' in which the author Muramatsu Shigekiyo uses a 2^{15} or 32768 sided polygon to calculate the value of the number to a great degree of accuracy.

Most mathematicians, usually samurai or the warrior class, who had received the 'Master of Mathematics' licenses visited rural schools to teach mathematics. It was in this milieu, isolated from the Western world and increasingly divergent from China, that *wasan* or Japanese mathematics arose. Traditional methods created the most striking results, and thus Japanese mathematics skirted the formalised 'calculus' but became no less ingenious and persisted to add to the plethora of traditional sacred mathematical patterns and designs, puzzles and solutions that bathe and glorify the mystical nature of the many realms of mathematics.

CHAPTER EIGHT

MYSTICAL NUMBERS

"Eko ham bahu syam (I am One, may I be many)."

—The Upanishads

Since antiquity particular numbers have been considered significant and imbued with mystical properties. What impact they have had in cultures and societies that identified them is uncertain, but it does seem plausible that many rituals and belief systems had been woven around these sacred numbers. Although there are innumerable examples of special numbers relating to specific religions and cultures and even scientific traditions, here we shall only consider a few notable ones that feature in a variety of contexts – including religions such as Hinduism to Sufi Islam and far eastern cultures.

The fundamental question that has always been the focus of my own quest regarding the significance of such numbers is: can there emerge a different kind of mathematics out of spiritual and 'organic' considerations? Arising out of this are further questions, such as, what might this *organic thinking* be? Galileo Galilei has been attributed as the father of modern science, and indeed it is remarkable that his mathematical insights were greatly influenced by the humanities and the arts, specifically by Dante's Divine Comedy. Drawing from this example, we might ask a daring question : can a 'mathematics' emerge from a more spiritual and perhaps God conscious approach to number? Further, can this mathematical thinking foster a closer observation of and immersion in nature? Expressed differently, can a 'mathematics' emerge from a more holistic approach to learning itself where the interplay between varied disciplines promotes a more layered understanding of mathematical ideas and expression? And could this interplay include a mystical and theological approach to arithmetic in general?

THE NUMBER 108

The number 108 is considered sacred in the Hindu, Jain and Buddhist traditions of the East. In Hinduism, the *Mukhya Shivangas* or attendants of Shiva are 108 in number which is symbolically expressed in the use of *malas* or beads of 108 for prayer and meditation. In Gaudiya Vaishnavism, Lord Krishna had 108 followers or *gopis* and the ritualistic recital of their names is accompanied by the counting of a 108 beaded *mala* during religious ceremonies. Even the Sanskrit alphabet has 54 letters which are half of 108. In the Sri Vaishnavite tradition, 108 *divya desams* or temples of Vishnu are venerated and there are 108 sacred places or *pithas*. In the Jain religious tradition, 108 features prominently in the total number of ways of *asrav* or Karma influx – 4 *kashays* (anger, pride, conceit, greed) x 3 *karanas* (mind, speech, bodily action) x 3 stages of planning (planning, procurement, commencement) x 3 ways of execution (own action, getting it done, supporting or approval of action).

In Buddhism, according to Bhante Gunaratana, the Theravada Buddhist monk, 108 is arrived at by multiplying the senses – smell, touch, taste, hearing, sight and consciousness (6) by whether they are painful, pleasant or neutral (3), and again by whether they are internally generated or occur externally (2) and further by past, present and future (3) giving us a total of 108 feelings – 6x3x2x3 = 108. Tibetan Buddhist *malas* or rosary beads are also typically 108 in number and less frequently 111 including the Guru beads, emulating the words of the Buddha called Kagyur in Tibetan presented in 108 volumes. Zen priests wear a ring of prayer beads called *zuzu* around their wrists which also add up to 108.

At the end of the year in Japan, a bell is chimed 108 times in Buddhist temples to finish the old year and welcome the new one. Each ring represents one of the 108 *bonno* or earthly temptations a person must overcome to achieve nirvana. Furthermore, in Hindu and Buddhist Astronomy, 108 features in the relation between the earth and the sun's diameter. The average distance between the earth and the sun is 108 times the sun's diameter. Also, the average distance from the earth to the moon is about 108 times the moon's diameter. This happy 'chance' allows the moon to appear the same size as the sun during eclipses.

THE NUMBER 137

The number 137 continues to ignite the imagination of many scientists, mystics and occultists alike, according to Professor Arthur I. Miller, professor of history and philosophy of science at the University College London. In his book '137: Jung, Pauli and the Pursuit of a Scientific Obsession,' published in 2010, he talks about how the physicist Wolfgang Pauli was fascinated with numbers starting from his days as a physics student under his mentor Arnold Sommerfeld.

It was Sommerfeld who discovered the extraordinary number 137 in the year 1915, while trying to solve the fine structure of spectral lines, the characteristic combination of wavelengths of light emitted and absorbed by each chemical element – the fingerprint or DNA, so to speak, of each wavelength of light. This number was dubbed the fine structure constant (in fact it closely equals 1/137 which is often referred to as 137 for convenience). This unique number or 'fingerprint' was central to the fundamental constants of nature, specific quantities believed to be invariable throughout the universe, quantities central to the general theory of relativity and quantum mechanics itself. The Nobel Laureate physicist Richard Feynman indicated how the universe around us appears to have a fundamental structure that can be glimpsed through special numbers (Paul Ratner, Big Think, 2018). He called it, "one of the greatest damn mysteries of physics: a magic number that comes to us with no understanding by man."

Over time the number 137 started acquiring mystical and symbolic status among scientists, psychoanalysts like

Karl Jung and musicians like Arnold Schoenberg, who were intentionally going against the grain through their musical eclecticism. This number also features heavily in the study and practice of the Jewish mystical tradition of Kabbalah.

THE NUMBER 7

In many of the world religions and traditions the number 7 bears a special significance. Even Pythagoreans interested in musical patterns were alerted to its distinctive qualities. There are seven distinct notes in the musical scale corresponding to the white notes on a piano. Counting from 1, the eighth note up the scale is the 'exceedingly harmonious octave,' which is how the name arose.

Genesis 1:1 is the first verse of the first chapter of the Book of Genesis in the Old Testament in the Bible and the opening of the first creation narrative. In the Masoretic text, the Aramaic text of the 24 verses of the Tanakh in Rabbinic Judaism, the verse consists of seven words. There are seven deadly sins in the Bible, along with seven virtues, and in the Book of Revelation, seven is the central figure of quantities: the seven spirits of God and the seven churches of Revelation to which the book is addressed. The old testament is sprinkled with references to seven; for instance, Noah is commanded to bring seven pairs of every clean animal onto the ark (Genesis 4:15) and there are seven days of Creation (Genesis 1) – God rested on and sanctified the seventh day (Sabbath). The New testament is also riddled with references to this number such as seven loaves multiplied into seven basket fulls of surplus (Matthew 15: 32-37).

The number seven is also mentioned in the Quran 28 times which itself is a multiple of seven. The 'basmala' or the phrase "*bi-smi llāhi r-raḥmāni r-raḥīm*" is also linked to the number 786 which is nearly a perfect square of 28. The 'basmala' is the Islamic phrase that means "In the name of God, the Most Gracious, the Most Merciful," and is recited before each chapter or *surah* of the Quran except the ninth. Numerologically, the phrase adds up to 786 in the Abjadi system and has acquired special significance in folk Islam and near eastern folk magic.

The first time the number seven is mentioned in the Quran is in the 196th verse (*Surah tal Baqra*) which is a product of 7 and 28. It is no surprise then, that the Arabic alphabet has 28 letters and forms the language in which the Quran was delivered. Furthermore, the number 196 has some interesting mathematical properties – it is believed to be a Lychrel number which is a natural number that cannot form a palindrome through the iterative process of repeatedly reversing its digits and adding the resulting numbers. This process itself is sometimes called the 196-algorithm alluding to the most famous number associated with this process – in general the reverse and add process that produces the sum of a number and the number formed by reversing the order of the digits. A few examples would be, say, 23 added to 32 to give 55, and 56+65=121, and 35+53=88. Here, 55, 121 and 88 are all palindromes. However, this does not happen with 196 for up to billions of digits, according to an ongoing distributed program run by Romain Dolbeau, and so far nothing that contradicts 196 being a Lychrel number has been found.

THE NUMBER 19

Another number of significance in the Islamic tradition is the number 19. In the Quran, the 74th chapter, 30th verse declares, 'over it are nineteen,' referring to the nineteen angels guarding hell.

The number 19 is also extant in the Baha'i religion. For example, the Baha'i calendar is based upon the year consisting of 19 months with each month having 19 days. Another occurrence of the number 19 is incidentally in the game of Go which was invented in China and is the oldest known strategy board game to be continuously played for the past 2500 years. The game is played in a 19 x 19 grid of black lines with 361 points.

THE NUMBER 8

In far eastern cultural traditions certain numbers are considered lucky or unlucky and determine how the dates for important occasions such as marriage are chosen. The number 8 has been regarded as the luckiest number among the Chinese, as the pronunciation of 8, 'ba', sounds similar to 'fa,' the word that indicates prosperity or fortune. Such is the positive association with this number that the 2008 Summer Olympic Games in Beijing began on the 8th of August, or 8-8-08 at 8 minutes and 8 seconds past 8pm local time. In Taoist culture, the number 8 is associated with wholeness and completeness. According to the *I Ching* or the Book of Changes, the *bagua* or the eight trigrams are the foundation for the revelation of everything. In Buddhism the eight fold path is a summary of the practices that leads to the liberation from *samsara* or the cycle of birth and rebirth.

Incidentally, the number 8, in mathematics is the first number that is neither prime nor semi-prime and is the only cube in the Virahanka-Fibonacci sequence. In computer science, the octal, 8-based system is used – a byte is a grouping of 8 bits.

THE NUMBER 28

The symbolism of 28 is fairly widespread across cultures and religious traditions. In the far east, King Sejong the Great of Korea (1397 - 1450) changed the course of Korean history by creating and introducing the Hangul script, the native phonetic writing system for the Korean language. Contemporary records such as the Veritable Records of King Sejong and Jong Inji's preface to the *Hunimjeongum Haerye* indicates that he invented the script himself. It is curious to note that when he first presented the Hangul alphabet, it contained 28 letters, the same as the Arabic alphabet – 28 being a perfect number with other fascinating properties that have been noted previously, such as 1+2+3+4+5+6+7=28. Aside from this property, 28 is the sum of the first primes (2+3+5+7+11) and non-primes (1+4+6+8+9). Prior to the creation of Hangul, the people of the Joseon dynasty at that time primarily wrote in classical Chinese alongside phonetic writing systems based on the Chinese script that predated the Hangul by several centuries. The fundamental differences between the Korean and Chinese languages and the large number of characters that needed to be learned meant that there was much difficulty in learning how to write using the Chinese characters for the vast masses of people. In order to solve this problem, King Sejong worked on and presented the unique Hangul alphabet for his people. The Korean TV miniseries 'Deep Rooted Tree,' 2011, based on Lee Jung-myun's novel of the same name, portrays the struggles of King Sejong as he works on his secret project for years – the creation of the Korean alphabet – which was considered a threat to the old Confucian order.

THE NUMBER 666

The natural number 666 in Revelations Chapter 13 of the Bible is called the number of the beast. "Here is wisdom. Let him that hath understanding calculate the number of the beast: for it is the number of a man; and his number is six hundred and threescore and six." Incidentally, it also happens to be a triangular number which means successive numbers, in this case up to 36, add up to 666. In other words, 1+2+3...+36=666.

Strangely enough, the number of the beast appears in a special palindromic prime number called Belphegor's prime which consists of the number 666 flanked on either side by one with thirteen zeros or 1000000000000066600000000000 0001. This unique number reads the same both backwards and forwards and is only divisible by itself and one. It was discovered by Harvey Dubner via the 'Dubner Cruncher' a computer program he and his son Robert developed in 1984. The name Belphegor refers to one of the Princes of Hell who was charged with helping people make ingenious inventions and discoveries.

6174 AND 495

Sometimes numbers behave in a highly unusual manner and leave mathematicians perplexed as to why they possess such quaint properties. One such number is 6174 also known as

Kaprekar's constant after the Indian mathematician D. R. Kaprekar. This number is notable because of the following rule:

- Take any four digit number, using at least two different digits (leading zeros are allowed)
- Arrange the digits in descending and then ascending order to get two four-digit numbers adding leading zeros if necessary.
- Subtract the smaller number from the bigger number
- Return to step 2 and repeat the process.

This routine outlined will always reach its fixed point 6174 in at most seven iterations. The only numbers for which the above routine does not hold are 'repdigits' such as 1111 which gives 0000 after a single iteration.

The same Kaprekar routine algorithm can be defined for three digit numbers as well. The number 495, which is just one removed from 496 (a perfect number), also gives the same result as 6174, upon following the outlined routine. There is no explanation as to why these numbers behave in the manner demonstrated. Maybe with the passage of time, these outwardly unassuming numbers will gather more significance.

THE NUMBER 5040

In a recent paper, 'The harmony of the cosmos, the soul, and society in Plato,' (2018), Joseph Milne delineates Plato's ordering of the *polis* of 5040 households by stating,

One remarkable way in which Plato conceives the proportionate ordering of the polis in the 'Laws' is to limit the

population to 5,040 households or extended families. The land for such a city should be large enough to support its population moderately, without excess, yet sufficient for defence against injustice from neighbouring cities, and strong enough to aid neighbours if they suffer injustices. The number 5,040 has exactly 60 divisors, counting itself and 1, and also is the sum of 42 consecutive primes. It therefore lends itself to complex proportionate divisions of functions of the population.

Expanding on the details of the nature of the number 5040, a few prominent features stand out, such as 5040 also being factorial 7, written 7! = 1x2x3x4x5x6x7. The number might also be written as 10x9x8x7=5040. Incidentally, this number is also divisible by 28, the perfect number that is repeatedly encountered in many contexts.

THE NUMBER 1729

A recent encounter I had with the number 1729 while watching the 2015 film 'The Man who knew Infinity,' based on the biography of the mathematical genius Srinivasa Ramanujan (1887-1920), made me reflect upon the hidden qualities of numbers that could be overlooked easily. Only to be revealed later upon either contemplation or serendipitous discovery. In a short sequence in the film, Ramanujan's friend and advisor G. H. Hardy casually mentions how he took a taxi with the number 1729 and found the number to be rather boring. 'Quite on the contrary,' responded Ramanujan and went on to add how it was a curiously interesting number, the smallest one that could be expressed as the sum of two cubes in two separate ways, thus, $1729 = 1^3 + 12^3$ and $10^3 + 9^3$.

I too was riveted by this otherwise ordinary looking number and wonder if it is too audacious of me to extend its properties to the realm of the mystical if not the conventionally religious. For instance, 1729 is also 19x91 where the number 19 (and its reflection 91) resurfaces – numbers notably significant in the Islamic, Kabbalah, Sufi and Baha'i traditions and folklore. Also the cubes that add up to 1729, are of 10 and 9, numbers adding up to 19. Finally, the other cubes that add up to 1729, the numbers 1 and 12 may also have a religious significance across many cultures. In Islam for instance, the '*kalima e tauheed*' or the insistence of the oneness of God '*la ilaha ill'Allah*' has 12 syllables in Arabic – twelve syllables asserting God's unity, signifying the numbers 12 and 1. What Ramanujan failed to specify but probably meant was that 1729 was the smallest number that was the sum of the cubes of *positive* integers in two separate ways. If negative integers are included then the smallest number that has the same property of being the sum of two cubes in two separate ways is 91, or $6^3 + (-5)^3 = 91$ which is also $4^3 + 3^3$. Again, 91 is the 'digits switch' of 19 and is a factor of 1729, evenly dividing this number into 19 parts. These occurrences may be brushed aside as mere coincidences but meditating on a deeper meaning and significance of such numbers leave us with profound queries about the nature of number itself. And the fundamental link they might have with law. These questions in themselves are philosophically outside the scope of this work, but do prompt us, nevertheless, to ponder upon them.

Ultimately, it does seem to me that the apparent *lack* in modern mathematics is that it does not pay attention to number in a more God-conscious way of understanding mathematics. This *lack* of not being attentive to number or the seeking of meaning of number, purely for itself, may leave something essential about mathematical understanding out. The modern way of thinking appears to underscore 'use' in the seeking of number and number systems. Nonetheless, numbers in themselves have great meaning and a failure to attend to this meaning seems to be a dire omission. Pierre Teilhard de Chardin (1881-1955) has spoken of the mental body of things – what is the earth developing towards? He meant not just *ideation*, but the meaning of things. Thus, what is the meaning of number?

What is the sacred meaning of number? Specifically we ask again, what do particular mystical numbers mean in an absolute sense? Maybe the qualities of certain numbers lend themselves to contemplation. A revisiting, noting and observing of the inner beauty of numbers, for their own sake, might lead to a newly forged relationship with mathematics overall. If nothing else, this re-imagined relationship might be the very journey we need to undertake in order to truly appreciate the elegance and beauty of the subject.

CHAPTER NINE

THE MAGIC OF POETRY IN NUMBER

Lousy at Math

Once a group of thieves stole a rare diamond
larger than two goose eggs.

Its value could have easily bought three thousand horses
and three thousand acres of the most
fertile land in Shiraz.

The thieves got drunk that night to celebrate their great haul,
but during the course of the evening the effects of the liquor,
and their mistrust of each other grew to such an extent they
decided to divide the stone into pieces.

Of course then the Priceless became lost.

Most everyone is lousy at math and does that to God –
dissects the Indivisible One, by thinking, by saying,
"This is my Beloved, he looks like this and
acts like that, how could that moron over there
really be God?"

—The Gift: Poems by Hafiz, translated by Daniel Ladinsky

Years ago my mother who was a professor of Urdu and Persian Literature made an unusual comparison. When I remarked that my subjects of choice, physics and mathematics, were so different and far removed from her world of Persian Sufi poetry, she rejoined, 'At the highest level, mathematics and poetry become one.' Never doubting the accuracy of her statement, I was inspired and propelled to search for beauty and meaning both in the mathematical and in the poetical. I steeped myself in the poetry of the mystics and the greats and ventured further into Sufi *ghazals* and *shayari*, enthralled by the rhythmic words as though they were embroidered on a page that stirred up deep emotions that I barely knew existed within me. Thus mobilised, I began my dual quest to draw inspiration from both mathematics and poetry, seeking connections in the rhythm and *taal* of the poetical with the basics and foundational blocks of the numerical.

The first time I understood that mathematics and poetry must be interlinked, in a way that clearly demonstrated the connections, was when I encountered the Italian poet Dante (b. 1245 B.C.E.) in the context of the physics of scaling laws in mathematics, through my college advisor, Mark Peterson, who happened to be an interdisciplinary scientist himself. The unexpected connection between a work of literature by

Dante, and the science of scaling laws, espoused by the father of modern science, Galileo Galilei, in the 1600s, astounded me. I learned that Galileo had published his 'Dialogues Concerning Two New Sciences,' on the bending and breaking of beams and had carefully summarised rigorous experimental investigations of motion, including the laws of falling bodies and the parabolic path of projectiles as a result of mixing two types of motion – constant speed and uniform acceleration. A problem that had occupied Galileo's mind for years was that of scaling laws, or the stability of structures when they are scaled up or down in size. In his book Galileo's Muse (2012), Mark Peterson argues that Galileo's insights on scaling were influenced by works of literature, namely that of Dante's in his Inferno where he describes the physicality of hell in careful detail. Ultimately, Galileo was able to see that Dante's Inferno was unsound structurally, as the size of hell could not be scaled up without altering the scaling laws themselves. One way to plainly see what is implicated by these mathematical scaling laws is to observe the bone structures of small and large animals. We find that we cannot merely scale up the animals without altering the width to height ratio of skeletal bones – in other words there is a limit to the size of animals. Moreover, the ant and the elephant have very different body proportions because of their vastly different sizes and masses – hence larger animals have thicker bones.

In a previous chapter relating to the Virahanka-Fibonacci series, we had delved into the relationship of this mathematical series to the beats of a tabla, a traditional Indian musical instrument. It is therefore not much of a stretch to expect mathematical patterns, which are rhythmic in their nature, to be observed in the beats of music, song and poetry. Alain Badiou (b. 1937), the celebrated contemporary philosopher and former Chair of Philosophy at the Ecole Normale Superieur, talks about how the prominent philosophers before Plato and Socrates, were essentially poets; the Greek poet-philosopher Parmenides (b. 515 B.C.E) being a prime example. Badiou, in his 2017 lecture delivered at New York University, highlights the tension between the demonstrative tendency and the seductive tendency of poetry (in philosophy). He further insists that the transition of the poetical nature of philosophy to its rational nature occurs with Parmenides. Here we shall briefly delve into a sampling of prominent cultures and traditions where poets and poetry were interwoven with mathematics. We shall touch on the works of the individuals who were both poet and mathematician – or whose mathematics or poetry was inspired by the other.

Historically Sanskrit poetry is speckled with mathematical structure, which particularly shines through the works of Gopala and Hemachandra in the 11th century. As mentioned in Chapter 5, Manjul Bhargava has examined this topic thoroughly and highlighted the connections between Sanskrit poetic verse and mathematics. Beside the connection between the syllables in Sanskrit poetry and the number of petals in flowers, another extraordinary discovery is the connection between mathematical binomial structure and the Sanskrit verses of Pingala (c. 200 B.C.E.). In his poetry, if you start with n syllables, k of length l and n-k syllables of length 2, then you can construct lines of different lengths. There are $^{n}C_{k}$ ways of organising these which is the binomial expansion coefficient. When Pingala considered how to deduce the relationships between these different length poetic sentences he realized that they could be arranged in a triangular form (called Mehuprastara in Sanskrit), which is otherwise identified as Pascal's triangle, named after the 18th century mathematician Blaise Pascal.

The Islamic world is also full of examples of the marriage of poetics and mathematical rhythm, and the inspiration of mathematics or science drawn from a poetical orientation. The most famous physician of the medieval Islamic world, Ibn Sina (b. 980 C.E.), anglicized as Avicenna was born in Bukhara, in Hamadan, Iran, and was a major contributor in the fields of Aristotelian philosophy and mathematics. His medical poem, *Al-urjuzah fi al-tibb*, was significant in its role as a transmitter of medical knowledge to medieval Europe. The poem consists of 1326 verses, meticulously classified as a poetic summary of his encyclopedic textbook, The Cannon of Medicine. These medical writings played a key role, through his poetry, in the transmission of medical knowledge to medieval Europe.

Arabic didactic poems used in the transmission of knowledge have two characteristic features. The poems in this genre are composed in the '*Rajaz*' metre whose pattern of syllabic repetitions produces a 'jingling sound that is particularly easy to remember.' In sharp contrast to mono-rhythmic arrangements of standard poems, the rhyme of the first verse in an *urjuzah* is not followed in the remaining verses of the poem. There is, however, a constant similar rhyming of the two hemistitches or half-lines in each verse.

The expansion, speed of transmission and dissemination of knowledge must owe a great debt to poetry which I suspect also shaped the nature of the development of the science being transported. The language in which information is put forward influences much of what is being said, contrary to the idea that 'data' or 'scientific information' is dispassionate and 'factual.'

Another famous Iranian historical figure, Omar Khayyam, was a mathematician, astronomer, philosopher and poet. He was born in 1048 C.E. in Nishapur, in Northeastern Persia. As a mathematician and astronomer, Khayyam addressed important problems with incisive subtlety, attending to the problem of cubics by using intersecting conics and thereby presenting a geometrical interpretation of the solution. He also designed the solar Jalili calendar that is still in use today, even after a millennium. During his own lifetime, Khayyam was well known as a mathematician and scholar. The famous 'Rubaiyat of Omar Khayyam,' composed of quatrains is attributed to him and was made famous in the western world by the English author Edward Fitzgerald through his 1859 translation of the work. There has been much dispute about the theological beliefs and philosophical ideas of Omar Khayyam; however, the following quatrain ascertains Khayyam's belief in the oneness and non-duality of God:

Gar gauhar-e-tayaatut na Suftam hargiz
Gard-e-gunah as cheura na Ruftam hargiz
Bain hamaa naumeed nayum As karamat
Zan roo ki eki ra na du na Guftam hargiz

Edward Heron-Allen's (1898) beautiful literal rendering of the Rubaiyat is:-

If I have never threaded the pearl of Thy service,
And if I have never wiped the dust of sin from my face
Nevertheless, I am not hopeless of Thy mercy,
For the reason that I never said that One was Two.

A simple and yet forceful expression of the Islamic Doctrine of the Unity of God has been suggested by Khayyam in this *rubaiyi.* The last line of the *rubaiyi* is a virtual translation of the '*Kalemai Tauheed,*' (unification of the oneness of God). Incidentally, the third line of the *rubaiyi* reiterates Omar's faith in the ultimate Mercy of God which according to Islamic tenets has been guaranteed to all, who like Omar, do not believe in his plurality. Remarkably, the concept of the

'One' in contrast to the idea of the many, has been deftly made – in a manner of speaking, the beginnings of the theology of Arithmetic – the idea of unity and in contrast, 'number.'

Another famous medieval figure, Nasir al Din al Tusi was born in the year 1201 in Tus, Iran, where his father was a jurist in the Twelver Imam school, the main sect of the Shia muslims. A man of immensely wide ranging interests, Tusi was an outstanding Persian philosopher, scientist and mathematician, making original contributions both in mathematics and astronomy. The famous al-Tusi couple describes a geometric construction for producing rectilinear motion from a point on one circle rolling inside another. He also made original contributions of planetary movements based on his research at the Maragheh observatory. Al-Tusi was also a well published author who wrote on prose and mysticism. A poem that is attributed to him is reproduced here:

Anyone

who knows, and knows that he knows,

makes the steed of intelligence leap over the vault of heaven.

Anyone

who does not know but knows that he does not know,

can bring his lame little donkey to the destination nonetheless.

Anyone

who does not know, and does not know that he does not know,
is stuck forever in double ignorance.

Moving on to western traditions, an almost mathematical structure and technical consolidation of poetic form, the Petrarchan sonnet, together with the Shakespearean sonnet, are other examples of the balancing of rhythmic order and the beauty of poetic word. The sonnet has persisted in its appeal for over five centuries in western literature, increasing its range from Petrarch's (1304 -1374) Calzione - in the form of the Petrarchan sonnets, a sequence of 317 poems addressed to Laura, his idealised beloved. The first eight lines of each sonnet called the octave, states a problem, asks a question or expresses an emotional tension and the last six lines, the sestet, then proceeds to resolve the problem, answer the question or relieve the tension.

An illuminating study by two Russian researchers of Aleksandr Sergeevich Pushkin's 'Eugene Onegin,' a novel written entirely in verse (1833, single volume), teases out linguistic features using fractal dimensional analysis and complexity theory. The arrangement of the words in the verses of Pushkin's work follow the pattern that came to be known as the Onegin stanza, a modification of the Petrarchan style. The authors, Leonid Zhigun and Abdul Bangura (2018), explore the idea of spectrum progressing "from more orderly to less orderly or to pure disorder reflecting the directness, immediacy, purity, simplicity, and spontaneity of Pushkin's style." The results display the key mechanism of general self-organizing systems in the celebrated long poem.

In a longstanding series of studies on the subject, Alan Green has decoded the work of William Shakespeare (1564 - 1616) revealing an exceptional mathematical mystery hidden within the title page of Shakespeare's sonnets. Green proceeds to demonstrate mathematical 'encoding' even in the physicality of the so called 'frontispiece' of his celebrated book of 154 sonnets. He has claimed that a dozen mathematical constants are encoded by the proportions of the Great Pyramid of Giza. Furthermore, he found that three primary units of measure, the meter, the foot and the cubit are all related, connected by the basic proportions of the Great Pyramid and that none of those units of measure are arbitrary. In a more recent study Green presents a sacred geometry code in the title page of the famous book of sonnets, which alludes to the geographic coordinates of the Great Pyramid at Giza.

In the far eastern traditions, the Haiku, an ancient Japanese form of poetry of three lines with five syllables in the first, seven in the second and five in the last, has endured through the passage of time. In the words of John Gough, of Deakin University, 'although brief, a haiku is like

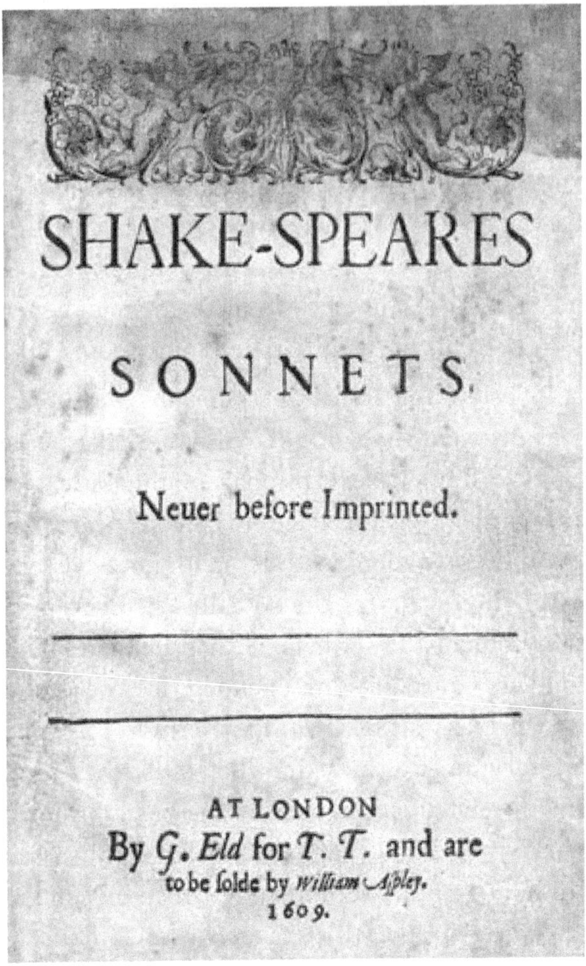

SHAKE-SPEARES

SONNETS.

Neuer before Imprinted.

AT LONDON

By G. Eld for T. T. and are
to be ſolde by William Aspley.
1609.

Title page of Shakespeare's Sonnets (1609)

a meditation on or observation of an experience, conveyed directly through objective images or sensory feelings with no personal judgement or analysis ... A haiku is a captivating photo of something in nature.' Presented below are some modern examples of attempts at capturing the Haiku poetic form with a mathematical theme as a topic within the vast space of mathematics, together with the formal structure of staying within the number and beat constraints of the original form. These appeared in the Journal of Humanistic Mathematics and have been collated in an article on the topic of 'mathematical haikus.' Three examples are presented here from the numerous poetical contributions.

Math in seventeen syllables: A folder in poetic form

Fibonacci

All Throughout Nature
The Fibonacci Sequence
Spiraling Outward

Transformed

Think of earthquake strength,
Musical notes and loudness,
And brightness of stars

Why fractions are Hard

Two variables -
not one. A relationship -
hence difficulty

We now arrive upon contemporary examples of the cross-fertilization of mathematical rhythm and poetry, demonstrating that the link is undeniable and continues to draw inspiration from poets and mathematicians alike as they begin to meld the lessons drawn from one area of study into the other, enhancing the nature of their works. Rosanna Warren (b. 1953), the well known American poet, has presented a clever take on prime numbers and their spacing in the poem called 'Number theory' reproduced here:

NUMBER THEORY

The four-and-a-half-foot black-backed rat snake swayed
up and across
the kitchen screen door, seeking
a way in. Encountering, instead,
our eyes, it slowly, deliberately, withdrew
to slide across the stone porch, over the wall, and along
the foundation, inspecting every crevice,
feeling, nosing, listening its way
toward a solution, which it found
around the corner, up the back flagstone steps,
where it squeezed its impossible length and girth, inch by
patterned inch, into the crack beneath the topmost slate. So
we know we're living with a patient companion,
like you, inquisitive.
You sit taut in your chair, whispering, as you probe
the gaps between prime numbers. Until infinity.
It's pattern you seek. The opening through which
your thought will glide suddenly into a lit space
and be at home. In a shaky house, where wasps gnaw the walls.

A rare analysis of how knowledge is acknowledged and synthesised into systems of thoughts and beliefs, 'Poetry and Mathematics,' by Scott Buchanan (1929) is a broadly relevant book on how both poetry and mathematics are mirrored in representations of human experience. While lecturing at the People's Institute in America, Buchanan explored the elements of these subjects and understood that the relationships between the words of poetry and the elements of mathematics deserve further probing. His later inquiry into the works of George and Mary Boole and other mathematicians led to the publishing of his findings about symbols in literature and measurement in the 1932 book, 'Symbolic Distance in Relation to Analogy and Fiction.'

A more recent study by Alice Major, 'Word Shapes and Rhymescapes: Translating Translation Symmetry into Music and Poetry,' (2013), creates the perfect template to explore the relevance of symmetry and symmetry-breaking that drives the creation of structure in so many things 'from the Big Bang to a human work of art.' She insists that it is easy to see how visual artists use symmetry transformations in creating their works. Elaborating her role as a poet, she feels she also uses these mechanisms as they are part of her toolkit in constructing words and creating poetic form.

From our meanderings into the intersections between number, structure, beat and rhythm, and their individual applications to both formal mathematics and poetry, we may surmise that the principles of one trickle onto the other, leading to the enhancement of both. Our instinctive inclination to poetic rhythm and beat, might be a reflection of our natural attraction to music and vibration and to order and rational mathematical thinking. We also notice that knowledge is easier to to transmit orally through time, through generations, in the form of poetry and verse, from epical *masnavis* to sonnet verses. This connection of verse and mathematics has been noticed through the earliest times – after all, Tartaglia (c.1500 - 1557) in his response to Cardano (1501 - 1576) on discovering the cubic solution responded in verse. Incidentally, Tartaglia happens to be one of my mathematical genealogy ancestors, if one traces all the way back from my PhD thesis advisor to his advisor and so on. The first few lines from the translation by Friedrich Katscher are:

When the cube with the cose beside it
Equates itself to some other whole number,
Find two others, of which it is the difference.

Hereafter you will consider this customarily
That their product always will be equal
To the third of the cube of the cose net.

Its general remainder then
Of their cube sides, well subtracted,
Will be the value of your principal unknown.

The quick tour through this chapter, of instances both historical and modern where mathematics and numerical precision or 'restrictions' play a role in the alignment of poetry, leaves us with more questions than a sense of closure. The question of the nature of poetry in relation to the mathematical is as profound as is the nature of music or pulse in the setting up of mathematical patterns. Who came first, the poet or the mathematician? One can only attempt to answer this question in the realm of poetical thought as it tries to organise and align itself to that which is beautiful.

Recently, I have joined a class in traditional Indian classical music where I am learning about *ragas* which are an 'array of melodic structures with musical motifs' providing a musical framework in which to improvise and create songs, hymns, or ballads. The first *raga* I have learned is called 'Raga Yaman,' which was memorised by the Delhi court poet Amir Khusro (1253 - 1325) when he was struck by a tune hummed by a Yemeni merchant. In another account, my classical music *Guruji* insists that the raga is derived from the words '*Ay man*,' loosely translated as 'dear heart,' a self-referential *raga*. In one of the traditional renditions of this raga, the first line, '*mora man harwa leeno daiya*,' literally means that my heart (love) is defeated (encased) only by the ocean. This raga is not only an ode to love, it is also a mathematical suggestion of the limits of infinity. Only the ocean can contain and thereby 'defeat' the intensity of love. The frequent interplay of poetry and pattern or poetry and the lofty concepts of infinity, zero or limits forms the rich ground for profound insight into the nature of learning itself.

CHAPTER TEN

MATHEMATICS AND ART

"Without dreams there is no art, no mathematics, no life."
—Michael Atiyah

Art and mathematics, both seemingly distinct and unrelated subjects, also appear to have a lot in common on closer inspection. They are both highly creative and abstract subjects and often concern shapes, symmetry, measurement and proportion; and they also involve the somewhat uncategorisable search for beauty. Even historically, mathematics and geometry in particular have edified art and vice versa. The boundaries of mathematics and art are oftentimes the most interesting areas of study or research. On the face of it, it might seem that mathematics as a subject has evolved in isolation – precision and order, repeatability and rigour being its hallmark. In contrast, art appears to arise from a much more intuitive space motivated by beauty. Nevertheless, I have always believed that this demarcation and indeed the allotment of camps for pupils in schools – the 'science type' of student and the 'arts type' – is indeed ill informed and never useful.

How often has art informed and thereby enhanced the study of mathematics and propagated its geometrical development? Two examples of relatively modern artists, Vincent Van Gogh (1853-1890) and M.C. Escher (1898-1972) serve to illustrate my point – that art must come first. It is possible to learn about geometrical patterns, and thereby develop mathematical intuition through art. Thus, art is the vision and mathematics is the laying out of the dimensions of that vision. The charting out of that vision of art is partly poetical and partly the rigours and requirements of order.

One may consider the astonishing painting of Vincent Van Gogh, 'The starry night,' (1889), as a case in point. Here, artistic intuition almost presupposes the solution of an otherwise intractable problem – that of mathematical turbulence. How Van Gogh might have anticipated the problem (and possible solution) of turbulence is baffling to say the least, but it does say a lot about mathematical, artistic and creative intuition. The intuitive process is peculiarly mysterious, laced with Grace. Thus, art might be the material manifestation of the beauty, rhythm and symmetry of ordered (mathematical) thought. As in the previous chapters, a smattering of both historical and modern examples of the intersection and interaction of these two apparently immiscible subjects may question the very basis for the frequent categorisation of students as 'arts' types and 'science' types early on in school.

The Starry Night, by Vincent Van Gogh, 1889.

A short historical survey of the evolution of mathematics and geometry in particular, leading to elaborate designs employed for art, both in the miniature and in the monumental, might provide valuable insight into the interconnectivity and intermingling of these areas since the beginning of artistic thought itself. We shall briefly look into the historical and cultural development of art in the context of how it was employed, through mathematics, both in relation to worship and dwelling, towards the expansion of civilisation and in directing its progress. Much evidence has been unearthed to establish that the marriage between art and mathematics had already taken place during the Indus Valley Civilization that dated back to the middle of the third millennium B.C.E. From the artistry of the potters and architects of the Indus age to the early renaissance 'mathematical' paintings of the Italian mathematician Piero della Francesca (b. 1415) and Albrecht Durer (b. 1471) to the more recent geometric pop art of Japan, we find mathematical inspiration within art and artistic inspiration for mathematics across many cultures and civilizations. This art might manifest in the form of patterns for the decoration of altars, in the motifs in temple walls, in the weaving of textiles and quilts and in the sculptures and paintings adorning town halls and galleries. As this topic is vast, we will have to rely on a few selected examples and stay with the profound question of the natural yearning for ordered beauty and even ask what it is. And indeed how it impacts our otherwise rather socially driven quest for mathematical understanding.

In the earliest Indian context, the geometric principles put forth in the Sulbasutras (800-500 B.C.E) were thought to be the inception of mathematics in the Indian subcontinent (Seidenberg, 1975; Staal, 2008). This collection of *sutras* codifies the directions for the construction of sacrificial fires, including rules for the complex configuration of ritual altars. The bird-shaped Agnicayan altar, consisting of five layers of two hundred bricks each, with the bricks being of square, rectangular or triangular shapes of various sizes, is considered by F. Staal to signal the beginning of geometry in South Asia (Staal, 2008). The Sulbasutras as a document clearly indicates knowledge of Pythagoras' theorem long before the time of Pythagoras, and also other mathematically advanced notions such as the estimation of the square root of 2 and the 'squaring of the circle.' Be that as it may, the discovery of the archaeological remnants of the Indus Valley civilization in parts of Pakistan and northwestern India over the course of the last century has revealed a deep understanding of geometry that predated the Sulbasutras by more than a thousand years. Space filling close fitted patterns found in seals such as the one excavated from Mohenjodaro with the basic motif shaped like a Japanese fan, (Shah and Parpola, 1991), indicate quite clearly the level of the knowledge of circular geometry, more advanced than previously assumed. Other instances where the geometrical knowledge of the people of the Indus valley appears to be considerably refined is the decoration on the surface of a jar from Mohenjo-daro showing intersecting circles creating a lattice in a space filling pattern. Also a seal impression and numerous ivory objects attest to the use of geometry in art. In Harappa, various geometrical motifs were also unearthed such as a star like geometrical design in ornaments.

Noting the complexity of the shapes used as art in urn pottery, vases and figurines, implies that the Indus culture may have been adept at accurately estimating the area of shapes enclosed by circular arcs. According to an informative study by the authors of 'In Square Circle: Geometric Knowledge of the Indus Civilization,' (2011), the origins of mathematics and geometry in the Indian subcontinent might actually date back to the Indus Valley civilization rather than the period marked by the Sulbasutras as previously conventionally believed. By closely focusing on the geometry of the design motifs rather than the architecture of the excavated sites on the macro scale, one can get a more accurate estimate of the level of sophistication of the knowledge of geometry that may have persisted in that early pre-Vedic period.

Nowhere is the relationship between geometrical order and artistic design more melded than in the art found in the Islamic world. From the lace-like *jaali* work or trellised decorations found in the middle east to the intricate Mughal designs of Fatehpur Sikri in north India, the harmonies of elaborately carved inlay-work and etched designs are indeed astounding and a vast subject in themselves that would require several tomes to do them justice. The unique methodology of patterning made by the technique called *alicatado de la Alhambra, Torre de la cautiva* was used, as the name suggests, in the decoration of the Alhambra in Granada, Spain. The ornate ceiling designs in the Topkapi palace in Istanbul, Turkey, are matched in their elegance and grandeur by the meticulous arrangements of patterned design-work in the ceilings of the Pink Mosque in Iran. Found liberally throughout the Islamic world, *jaalis* and lattices are essential to the expression and articulation of architectural space. Issam El-Said and Ayse Parman have written an inspiring and informative book illustrating the link between geometry, order and pattern in the Islamic world called 'Geometric concepts in Islamic Art,' (1976).

The extraordinary symmetries one finds in the Mughal art of the Agra fort and the intricate striking patterns of Fatehpur Sikri, built by the Mughal emperor Akbar (b.1542), have an unexpected ally in the order and precision of group theory, a relatively modern area of mathematics that studies algebraic structures and systems consisting of sets of elements and a binary operation that can be applied to two elements of that set which together satisfy certain axioms and basic properties. The fact that Islamic art is non-figurative, allowed artists to explore the depths of geometric arrangement and design and the symmetrical sophistication and possibilities of expression using merely straight lines and circles. It would be a question worth pondering whether the idea behind group theory was motivated, in a sense, by the study of rotational and translational order in art and design; in the nested tessellation of patterns with origins in Islamic art or even in pre-Islamic, ancient or pre-historic self-similar repetitions of form.

The Pink Mosque, Shiraz, Iran

It struck me as highly providential when I came across a recent treatise on the striking engagement between portable arts and architecture in the medieval Islamic world by Margaret Graves, 'Arts of Allusion: objects, ornaments and architecture in medieval Islam,' (2018). This book argues for material objects and craftsmanship as a vital component of the intellectual culture of the medieval Islamic world. It attends to the connections between the material arts of the hand in miniature form to an expanded, larger manifestation such as a building. Portable objects like vases, incense burners, urns and lanterns, are reflected in the very large – or in the monumental – as the author points out, via architecture. As both these forms are three dimensional, Graves singles out some examples of objects that particularly resonate with the building of architectural monuments, as though these objects in miniature form were templates or blueprints of the larger realisations. She also asserts through her engagement with these artifacts that making is a form of thinking. In the context of this chapter I wish to extend this idea to include the making of these artistic ornaments, artifacts and objects, as a form of 'doing' mathematics via the hand. Much like rhythm and thinking are embedded in the pianists fingers, 'making' or more strictly, the '*poesis*,' of these ornaments also has mathematical geometry embedded in it. Ceramics, metal carvings, jewellery, and other objects of everyday use have been created in a poetical way from the senses and the brain, body, and hand. Further to what Graves suggests in her book, I would venture to bring in the mathematical as arising from these very tactile senses of the body – the creation of this kind of art as an instinct in a way, that eventually directs the mathematics which then reshapes the very art that was the original inspiration.

It is believed that Arab mathematics and Renaissance art were indelibly linked in a way that created great works of painting, sculpture and architecture in the period from the

14th to the 17th century. In a current article published online in *www.tandfonline.com* (2020) Hans Belting examines the transference of Arabic visual theory to the West, based on the book on visual optics by Ibn Al-Haytham (c. 965-c.1040), who was also known as Alhazen in the West. The article speaks of a double perspective, one of visual theory (Arab optics) and the other of pictorial theory (Renaissance perspective in art) and identifies the importance of Arab optics that led to realism in Renaissance art. Whatever the source or inspiration behind the burgeoning of highly awe-inspiring works of the Renaissance artists, the mathematics that followed created fertile ground for what we now know as modern mathematics. The art of that period frequently used both sacred geometry and linear perspective in the execution of the works, as we have noted earlier in chapter 7. The fundamental problem of coping with the accurate representation of our mind's three dimensional perceived reality gave rise to the illusionistic tradition of creating such an effect on two dimensional surfaces using techniques that were surely borrowed in part from mathematical principles. J. V. Field's 'The Invention of Infinity,' is preoccupied with the journey of arriving at the solution of the representation problem by artists and mathematicians of the Italian Renaissance. The area of projective geometry that naturally arose from attending to the purely mathematical aspects of this problem over the next two hundred years is deftly illustrated in this illuminating book.

One artist who stands out for his elaborate use of mathematical techniques is the Italian mathematician and early Renaissance painter Piero della Francesca (c.

1415 - 1492). To his contemporaries he was known as a mathematician and geometer and was later noted for the sublime humanism in his paintings and the liberal utilisation of geometric forms and perspective. Most well known among his other famous works, is his cycle of frescoes '*The history of the true cross*' in the church of San Francesco in the town of Arezzo. In a response to what came first, the art or the mathematics, Piero della Francesa's paintings in his book 'De Prospectiva Pingendi,' certainly suggest that he was indeed mathematically gifted and it very well might be that his love of mathematics arose from his artistic bent.

Another artist and printmaker Albrecht Durer who was born in 1471 in Germany was lauded for his extensive body of work that include altarpieces and religious works, portraits and copper engravings. His celebrated creation 'Melancolia 1' is widely considered to be the epitome of classical printmaking. As he was a practising mathematician, many of his paintings continue to fascinate artists and mathematicians alike. The images in Melancolia 1, which include 'Durer's polyhedra,' are particularly relevant in modern mathematical applications pertaining to the study of quasi crystals whose shapes are never exactly repeated but form fascinating structures nonetheless. For Durer, in a sense, to 'predict' these structures only studied to their full extent in more recent times does suggest that the source of the intuitive process in art is both mysterious and intriguing to say the least.

In modern mathematical history, the extraordinary works of M.C. Escher, who was born in 1898 in Germany, may be among the most profound examples of art and mathematics coming together in a manner that suggests the amalgamation of the artistic and the mathematical. It is hard to distinguish where one begins and the other ends. Escher had mastered the art of tessellation, which is the assembling of regular patterns that fit together perfectly and divide the plane without leaving any gaps. The Moorish designs from the Alhambra, when he visited the grand building in Spain in his youth, inspired him to such an extent that he became riveted by the possibilities of alignment, order, structure and symmetry in such a manner that his drawings anticipated deep features of modern cosmology and dimensional topology. Although he had no formal training in mathematics and always considered himself lagging behind the subject in school, his preoccupation with art had geometrical aspects even in his earlier works. After his meetings with mathematicians, especially after his interaction with the contemporary mathematical physicist Roger Penrose (b. 1931), his mind-bending art began reflecting their fruitful collaborations. A few of Escher's iconic sketches have their origin in Penrose's own mathematical sketches and ideas. The result of their interactions also lead to some fortuitous mathematical developments in Penrose's own works. Here was a historically recent example of mathematics being inspired by art. And this art itself could not have arisen from any formalised training. Later works by Escher, after his mathematical collaborations with Penrose and Harold Coxeter (1907-2003), exploited the idea of two dimensional geometry sitting within a space that was different from the standard Euclidean one. Escher tried to illustrate the concept of infinity by using curved squares and curved triangles repeatedly to build his artwork of non-overlapping fish that get smaller and smaller as they move from the centre towards the boundary. This particular work done in 1958 is called 'Circle Limit 111,' and is mathematically known as Poincare's disk as a representation of 'hyperbolic space' in non-Euclidean geometry. As it turns out, Escher had rather accurately displayed what the 'infinite' universe might look like in curved space.

This book has been partly written in the university town of Shantiniketan in Bengal, the home of poet and polymath Rabindranath Tagore (1861 - 1941) who won the Nobel Prize for Literature in 1913. Tagore introduced the Javanese batik technique of wax resistant dyeing to textile production via his art school at Viswa Bharati University which he founded in 1921. This ancient form of pattern creation using methods like stamping, block printing and wax dyeing may have originated in Asia and spread to the Malay archipelago through trading routes. Although now practised all over the world, this style of fabric printing continues to flourish in Indonesia, Malaysia, Japan and in its culturally adapted Indian form in Gujarat, Kutch and Shantiniketan, as both a means of livelihood for its practitioners, and as a cultural symbol of wide use.

As I visited the local workshops in Tagore's town, I became fascinated by the alternating use of molten wax and coloured dye in the creation of batik artwork. The portions of cloth covered in wax resist the dye and thereby retain the original colour. Painting patterns into the fabric before the batik process is completed allows the artist complete freedom in painting. Another form of batik using blocks with engraved designs is also frequently employed. The block printing of material using wooden blocks, much like the tessellated tiles of Islamic art and Escher's engravings, form unique patterns with mathematical precision.

A characteristic 'cracking' develops on the surface of the wax which allows small amounts of the dye to seep in. These fine cracks give the fabric an uncommon look – similar to fine capillaries traveling along the material mimicking fluid flow in minutely thin channels. This technique could be used in decorating *saris* and other clothing or home-ware like tablecloths. In Shantiniketan, the use of vegetable dyes has been revived, both for cutting costs and as an environmentally friendly alternative to chemical dyes. Most fascinating for me was to observe the naturalisation process of adapting the batik patterns and motifs to reflect the local culture and traditions of the arts, dance and agriculture. It is amazing to see the role of mathematical symmetry, both translational and rotational, in the assembly of many of the batik patterns, especially in the block printed motifs that exploit ideas of self-similarity. The presence of symmetry groups in the patterns appear to have symbolic meaning in the transmission of stories and histories. As I participated in the actual dyeing and making process of an elaborate patterned sari, I began to realise what an effective mathematical tool this was for the study of geometrical ideas in mathematics and how it can be easily harnessed in teaching young children simple geometrical concepts of rotation, translation and reflection in a visceral and more tangible way using the hand. When these children encounter abstract algebra and group theory at a later stage, it would come more naturally to them; having had a practical introduction to the elements and properties discussed in group theory via the practise of batik art.

The possibilities for the exploration of the mathematical elements in artistic endeavors seem limitless, based in ancient traditions that range from the carpet weaving arts and crafts of the caravan route of the middle east to the artistic batik patterns of Java to the origami paper folding techniques of Japan. Artistic and mathematical expression appear to be intertwined throughout history. A more subtle reflection of the artistic process also suggests an unexpected overlap into the 'making' of mathematics, as though we must doubly attend to the Greek roots of the word mathematics, or '*mathematikos*' which means *all that is already known* and poetry or '*poesis*' which is *making*. Perhaps we are born knowing the mathematical, and therefore a complete unraveling of the 'pre-knowing' must take place by a more natural introduction to it through artistic, poetical and musical exposure. It may very well turn out that the early resistance that is often found in the classroom setting will transform into a ready 'receiving' that is more holistic when exposed to nature, and in its context, varied arts and crafts. A more 'humanist' exposure to learning would provide the correct foundation for truly integrated learning and most of all will serve in our search and thirst for beauty which is indeed the greatest motivator of all.

CHAPTER ELEVEN

LEISURE AND FRIENDSHIP IN MATHEMATICS

"Even after all this time the sun never says to the earth, 'You owe me.' Look what happens with a love like that. It lights the whole sky."

—Hafiz Shirazi

Ordinarily, *friendship* or *leisure* are not the words we would expect to be associated with mathematics. Nevertheless, in all my days as an aspiring mathematician and scientist, I had found both these qualities to be central to my learning journey. As previously mentioned in chapter 2, the meaning of 'school' is derived from the Greek '*skhole*' which indicates a place of leisure – 'learning through leisure' being the highlighted idea. Curiously, I have found that friendship is another essential ingredient in the flowering of mathematical ideas through exchange. I present here, one such remarkable experience in Turkey, relayed through a personal lens, that might serve to explain how powerful leisure, friendship and freedom can be in the development of brand new mathematical ideas along with the rediscovery of old ones.

The town of Şirince

NESIN MATEMATIK KÖYÜ – A JOURNEY THROUGH GENIUS

Not far from the ancient Turkish town of Selçuk near Ephesus, lies the sleepy hamlet of Şirince (pronounced She-rin-jay). True to its name which means 'pleasant,' the town is utterly charming and beautiful, with old Ottoman houses cradled between rugged hills that appear almost incidental to the rest of the landscape. One could spend hours peeking into its strikingly lovely boutique hotels, coyly parked on various hillsides, or meandering through undulating streets dotted with shops selling wine, olive oil, carpets and crocheted table cloths, among other traditional wares. Apart from restaurants, cafes and shops proclaiming the best local *gözleme*, a sort of pancake like sweetmeat, Şirince is also home to the world's first 'Mathematics Village.' Founded by Professor Ali Nesin, Head of the Mathematics department at Istanbul Bilgi University, and named after his father, the poet Aziz Nesin, it is indeed an oddity in today's era of degrees, diplomas and certificates because it offers none of these. Students come from all around Turkey to learn mathematics from instructors offering free lessons: academics, researchers, teachers from all over the globe, uniting in their efforts to unveil the secrets of the subject. This is not only a unique experiment, probably the first of its kind anywhere, but a trail-blazing enterprise of momentous significance and a true eye-opener for educationalists in general. Students who have had the privilege to study here have left with not only a deepened love of the subject but a sense of camaraderie, an awakening of the mind and a spirit of wonderful mutual generosity. All this has been achieved in spite of difficulties created by an unsympathetic bureaucracy ever since its inception in 2007.

The Mathematics village was built in the manner of the ancients, with stone-walled houses, supported by wooden beams and caked with mud. Students in the previous years had helped in its construction and it initially included accommodation for the 60 odd students who assembled for part of the summer every year. Two separate *hammams* with blue domes provided afternoon respite during the scorching summer months. Not every student attended every class or course offered; they were chosen according to individual levels and areas of interest. Most of the lectures were conducted by practising mathematicians, although some advanced students also delivered lectures or seminars on selected topics. In the evenings the students studied by themselves or in groups, seeking advice on difficult or unsolvable problems, consulting with one another, enjoying each others' company until the wee hours. I was lucky to join in on the regular evening musical soirees, where the lilting voices of some of my students, singing folk songs on their *balamas*, a traditional lute-like instrument, filled the air, creating an atmosphere of magical intensity, each song seamlessly blending on to the next.

Even the classical architecture of the village seemed to inspire mathematical contemplation. Trees grew from the floors of the classrooms, soaring through the thatched roofs. I taught in an Ionian style outdoor theatre with seats of local stone draped with Turkish kilims while cloth pennants fluttered in the breeze overhead. The whole village was laid out in a way that encouraged constant movement; there were no dead-ends. Rather, each area of the village flowed into the next in a most ingenious way, like a scientist's mind sailing through a sea of mathematical forms and relations. There was no final harbour except truth itself.

Our room had a welcoming feel to it with an Ottoman style balcony at the entrance. The ceilings were high and the walls were thick to provide insulation from the oppressive summer heat. Turkish *kilims* and traditional-styled cushions added to the charm of the room, which was also adorned with a semi-nude portrait of a *Da Vincinianesque* woman. As it turned out, this oil painting was a work by Ali Nesin himself. Apart from mathematics and painting, he also demonstrated his great skills as a leader and administrator, for the place was run with remarkable efficiency.

The village had a certain magical ambience. The dining hall of the complex broadcasts this especially with its exquisite natural colors: olive, maroon, beige, and the colour of candle light. Trees grew through the thatched ceiling and one dined overlooking the valley of olive trees spread out below. The timeless beauty of the architecture, with stone, straw, clay and beams of solid wood – all procured locally – was unlike anything I had seen. Many students had physically helped in the construction of this complex, a true labour of love.

The classrooms were built around flowering trees of the olive groves, some rising from the ground to the ceiling, silent, unobtrusive witnesses attending the lectures. There was plenty of foliage providing a serene interplay of light and shade through the leaves and branches. Apart from regular classrooms, students could choose to complete a particularly obdurate proof sitting in one of the shaded coves. Or their romantic trysts with mathematics could continue more casually by the swings. The interior paths were so laid out as to continue from one end of the village campus to the other with no discontinuities, quite like a smooth mathematical function that reveals the nature of the problem.

The kitchen and dining hall, typically the hub of any institution, were not only the site for frenzied chatter, exchange of ideas, discussions, and problem-solving, but also a place where minds melded in creative learning and mathematical debate. We were spoiled with the most delectable Turkish dishes like eggplants stuffed with rice, onions, tomatoes, and dribbled with lemon juice. It was an unusual culinary haven, painstakingly created by the talented chef, an attractively stout woman who was an erstwhile bus driver, and the many students under her stewardship. I was introduced to stuffed vine leaves, *baba ganoush*, *lokma*: deep fried fritters with honey syrup and stuffed artichokes in olive oil, among an endless sea of gratifying home-recipes made with local ingredients and traditional spices.

The informal nature of the whole enterprise belied the mathematical rigour with which each topic was taught and studied. Some of the advanced topics were in areas as diverse as Fractal Geometry and Algebraic Coding Theory, cutting edge courses that any university would be proud to include in its curriculum. Several of the topics were those that students would not ordinarily encounter in their regular syllabus – topics especially chosen to display and introduce the range of mathematical possibilities. I taught Classical Mechanics for two successive summers there – a short course that covered the journey from Newton to Lagrange and Hamilton, passing over reluctantly to quantum mechanics, its natural successor.

Along with the rigours of modern mathematics comes the poetry of new discovery and infinite learning. I was routinely impressed with the passion and seriousness with which the students approached their classes. When I briefly touched on the calculus of variations assigning a notoriously difficult version of the *brachistochrone* problem, I found that the more eager members of the audience had not only finished the question, but had found alternative solutions and written them all out in cursively decorative handwriting suggesting several committed hours of labour during long nights. Yet, despite the seeming inaccessibility of higher mathematics and the kind of dedication its mastery entails, the atmosphere of the whole village was warm and light-hearted – one could frequently hear the friendly background banter of students joking and chatting, weaving through the interconnected pathways that linked one classroom or amphitheatre to another clearing or veranda, the venue for their next class.

Although my husband, who accompanied me on this trip, did not have an official teaching post in the first year, he easily wove his way into the spirit of the village. Impromptu philosophical conversations with several of the professors gradually evolved into more in-depth conversations about the history and development of mathematics from the philosopher's point of view. This caused enough excitement in the village that he was invited to teach a course on the Philosophy of Mathematics the following year, which turned out to be highly popular. He also got on very well with the students, volunteering to teach them English during the mid-day siesta hours and early morning hatha yoga on a nearby mountain top at dawn.

This easy spirit of adaptation was one of the main qualities that made the Mathematics Village such a wonderful place. The village somehow seemed to bring out the best qualities in everyone: selflessness, care, vigour, a ceaseless spirit of learning and deep curiosity. It made me wonder how all these qualities were achieved without the conscious striving for them that I often encounter in most institutions. The village was a community in the truest sense, a place where all ages were welcome, from families and older relatives to the youngest infant. They added to the tapestry of the place, residing for some weeks or months as accompanying guests. The cook charmed us every day with her culinary wizardry. She even started cooking primarily vegetarian dishes as a gesture of hospitality towards us, the new guests. The students helped in the kitchen and participated in the daily chores, through a rota system that *inter alia* involved wiping tables, the upkeep of the dining facilities, dish washing and peeling vegetables. Much to my surprise, they fully immersed themselves in the general care of the place that included attending to the small vegetable garden with enthusiasm and gusto.

My general teaching philosophy and outlook is that the teacher is also a student. I often learn about effective teaching and learning through my students who offer me suggestions and tips, both directly and indirectly, on how to progress with my lectures. My goal as a teacher of Mathematics has always been to express in the simplest terms ideas or concepts that otherwise seem cloaked in difficulty. Therefore, the art of teaching a notoriously difficult subject like Mathematics, particularly in its more advanced forms, involves a delving together to uncover the symmetries and underlying principles of the subject. Thus, together with the students, we discovered how Newtonian Mathematical Physics, which I introduced at the beginning of my lectures, gives rise to a more powerful approach introduced by

Lagrange, Hamilton and Jacobi, mathematicians who paved the way for the beginnings of Quantum Mechanics as well as a more complete understanding of the laws of Physics and the nature of Matter and Energy. The lectures, mainly conducted on a blackboard, using chalk and duster, lasted for about two hours each including the question and answer sessions and systematically traversed the journey from Newton's Laws to the beginnings of Quantum Mechanics.

Another fortunate encounter during my stay at Nesin Matematik Köyü was with Alexandre Borovik, a professor of mathematics I had first met during my doctoral studies at the University of Manchester. He pointed out that hidden structures and concepts of Elementary Mathematics which frequently remain unnoticed, nevertheless, significantly influence students' perception of mathematics. As an outcome of our breakfast conversations, he sent me a chapter of his then new book manuscript, Shadows of the Truth: Metamathematics of Elementary Mathematics, for my feedback and suggestions. While presenting the theme of his book, Professor Borovik extensively analyses real life experiences and episodes relating to the mathematical encounters children have when they are first introduced to the subject. In studying the formation of mathematical concepts in a child, he gains insight into the interplay of mathematical structures within Mathematics itself. He suggests that a philosophically inclined reader would notice a parallel with Plato's Allegory of the Cave. The children in his book see shadows of the Truth and sometimes find themselves in a psychological trap because their teachers and other adults around them are blinded both to Truth and to the surrounding shadows. I was fascinated to rediscover the hidden mysteries of Elementary Mathematics, realising that as a teacher one had to possess substantial insights into the structure of the subject even when teaching the basics. I enthusiastically wrote back to the professor offering a few tips on dimensional analysis and its history and a few other suggestions.

That it was a unique place came as no surprise to me, but I certainly was not prepared for this collage of natural and artistic beauty, right from the *hammams* where hot water was generated by the burning of olive pits, to the Ionian amphitheatre, a reproduction of an ancient open-air classroom of the Hellenistic period. The serenity of the village was punctuated by the chirping of crickets, in some ways reminding us of the meaning of nature's silence even with a welcome, but continuous, background noise. There was no TV or radio, in the days before the smartphone, to serve as distractions, otherwise routine presences in our technology and mass entertainment driven societies, forcing us to consider other means of unwinding after a full day. Students would bring out their *balamas* and other traditional instruments singing with full-throated, sonorous voices. A few carried an impressive repertoire of hundreds of traditional folk tunes, each song melding seamlessly into the next. One could spot batches of students and professors scattered through the dining hall forming small groups – some animatedly participating in impromptu musical soirees, some deliberating over a mathematical quandary, some playing cards, chess or bridge and some sitting still and contemplating the richness of the cloudless night-sky with a halcyon view of thousands of stars and an endless dark ether.

As I look at the photographs from those memorable summers, a collage of images float before my eyes: a professor finishing the initial draft of his book on the underlying principles of elementary mathematics; evening discussions ranging from advanced physics to the shibboleths of Sufi poetry; students tuning their traditional musical instruments; a blackboard covered with writing from an advanced algebraic topology class, and a spread of ambrosial food!

There was no talk of exams. They were non-essential to the purpose of the place and yet doing well at university or graduate school was an effortless by-product of all this effort. For those that set their foot inside Nesin Matematik Köyü, learning mathematics was uniquely exhilarating, an avenue to further goals, even if not directly linked to mathematics. A natural question that suggests itself is: what does an endeavour like the mathematics village mean for mathematics education in general? Does mathematics reveal its nature more fully when studied among friends purely for its own sake without any sense of comparison or the usual academic competition prevalent in typical universities and colleges? Is this unique village an indicator of how learning should ultimately be unfettered, with a genuine spirit of sharing through friendship and leisure? Perhaps it is merely the fulfilment of a dream conjured up by a few. Or perhaps, in the words of Ali Nesin, it is a beacon of hope for the coming generation.

CHAPTER TWELVE

SEARCHING FOR A NEW PARADIGM FOR HOLISTIC LEARNING

What is right education? This has been my lifelong quest in response to the rat-racc competition culture of the modern technological era. This question diverted my career from trying to become a mainstream scientist and professor of mathematics and physics to engaging as a teacher of high school aged pupils and young adults at an unusual and extraordinary school in Hampshire, UK, called Brockwood Park. An atmosphere of inquiry and questioning authority irradiates this unique centre for learning in the idyllic downs of Hampshire county. Gently rolling hills and vast open spaces with tucked away groves and invitingly safe forests set the stage for what would comprise some of the most mind broadening years of my life. I came to the realisation that it is only when the whole person is educated, in every fibre of their being, without a sense of partitioning between subjects, that true understanding can take place. This kind of learning must occur without the pressures of performance or competition, or despite them. In my own personal experience I have found that making connections between subjects and topics is illustrative for even the most specialised fields. These connections become impossible to properly articulate if there are too many students for each teacher. I have come to understand that low student to teacher ratios and beautiful location are the two most important elements for achieving genuine holistic learning. Realistically, this only happens when parents and teachers get together and create it; otherwise "education for profit" takes over, class sizes balloon, and the desire for rental income interrupts the natural beauty that nurtures holistic learning.

The philosopher Jiddu Krishnamurti (1895-1986) who founded Brockwood Park School in 1969, has said "When you look around you, not so much in the human world as in nature, in the heavens, you see an extraordinary sense of order, balance and harmony. Every tree and flower has its own order, its own beauty; every hilltop and every valley has a sense of its own rhythm and stability. Though man tries to control the rivers and pollutes their waters, they have their own flow, their own far-reaching movement. Apart from man, in the seas, in the air and the vast expanse of the heavens there is an extraordinary sense of purity and orderly existence. Though the fox kills the chicken, and the bigger animals live on the little animals, what appears to be cruelty is a design of order in this universe, except for man. When man doesn't interfere, there is great beauty of balance and harmony."

So education "when man does not interfere," has been my lifelong quest and this inquiry has taken me to a variety of places where the search for holism was in full swing though each place had its own speciality and unique set of challenges. I talked about the Turkish mathematics village in the previous chapter and now I wish to give a few more examples, from experience, of how holistic learning can happen in surprising ways. The first is a teacher's academy, held in the Himalayas in 2020 shortly after the severe Covid-19 lockdown in India. The second instance is a gathering of students from Brockwood Park School who came to India several years ago for an intensive study session in what came to be known as the 'Calcutta Academy,' a nod to the original space for free discourse and enquiry: Plato's academy. The third is the novel learning that has bloomed at a homestay and organic farm in a rural Indian village adjacent to the university town of Shantiniketan in West Bengal.

My family and I have been living in the eastern Himalayas for a decade. This area happens to be among the few remaining bio-diverse zones in India and is home to the broad billed warbler, the striped laughing thrush and the snow pigeon, birds that would delight any contemplative birdwatcher. Full of cascading waterfalls and natural springs, exotic orchids encased in translucent sheaths bloom in high-altitude thick broadleaved forests in this region of overwhelming beauty. We live with the Lepcha people, the inhabitants of the Kanchenjunga region since time immemorial. The year 2020 brought life altering changes to many across the world, including this Himalayan community. Quite by accident, during the period of sudden lockdowns in March of 2020, we found ourselves seeking shelter at the *Centre for Mountain Dynamics* in the Himalayan district of Kalimpong since our home in Sikkim was closed to us. As it turned out, this unexpectedly long stay of more than eight months gave us substantial insights on how *the right place* is also very important for right education. Contrary to the Cartesian view of looking upon all location as equivalent grid points, the impact of place as a central element in holistic learning cannot be over-emphasised.

The Centre for Mountain Dynamics (CMD) lies on a large piece of ancestral land, sandwiched by high mountain ranges and bordered by the lazily meandering Relli river that waits for the monsoons to swell to its full potential. For two decades, the CMD institute presented educational and cultural opportunities to the Lepcha youths including a sense of communal identity and political recognition. The flourishing of CMD lead to the opening of a school for Lepcha children in 2018 called The Mayel Lyang Academy, in itself an intriguing case study in long term self-sustainability and adaptable 'light-footed' education. I say light-footed because many of the structures that were constructed by way of lecture halls and classrooms, were built at minimal cost in the traditional style of crisscrossed bamboo-weave building, causing the least damage to these sensitive Himalayan ground rocks of dolomite, limestone and shale. The key to effective learning has been small class sizes, essential for the building of right relationship, which in itself is the best facilitator for the blossoming of every pupil. It is indeed rather charming that effectively conveying the essential principles of mathematics requires nothing more than paper and pencil and a good teacher.

Since its foundation in 2018, the twin principles of the Mayel Lyang Academy have been responsibility and inquiry. Teachers are encouraged to be creative using materials and non-structured learning in the form of real life examples. Each student receives individual attention aplenty which thereby fosters creativity. The small class sizes lead to a spirit of relaxation, promoting easy dialogue. Consequently, the imagination is continuously provoked in the individual. Forming connections from one topic to another seemingly unrelated one can provide linkages in the students' minds. A topic in mathematics related to numbers can bring about a connection in an allied subject – not just in science but in geometric design, art or music.

The students and parents together are involved in this broad educational purpose and they all participate in learning – via the children – granting the family an inroad into inquiry, learning and a broadening of vistas. The segregation and streamlining of age groups and the keeping out of parents and the wider community often takes away from the learning process. Especially on celebratory days including festivals, parents and other locals would gather in small batches – their indirect participation in the business of the school adding to the collective atmosphere of integrated community learning. The ethos and philosophy of the Mayel Lyang Academy help to retain *identity* and *spirit* and also promote the traditional foods and attire, together with the language and the endangered Rongrit script.

After the first lockdown of 2020 was announced in India, with barely a few hours notice, most income for the people of the Kalimpong district was abruptly curtailed. When the world was in panic, CMD quickly rose to the occasion and began cutting costs and generating revenue for its general upkeep by planting fresh batches of seasonal vegetables along with organic brown rice, and pairing up with an online food distribution network for local delicacies. The sheer resilience that CMD displayed in quickly adapting to the exceptional circumstances was exemplary and a suitable case study in itself for the long term versatility of such endeavours.

Lessons shifting online with limited face to face contact and a partial ban on get-togethers has put tremendous reliance on the use of smartphones and computers nearly everywhere. Many of the students of Mayel Lyang Academy live in remote regions of the mountains and are from financially deprived backgrounds. Several of them do not own smartphones or computers and must therefore rely on their books alone. Here too, the school hastily adapted to the fresh modes of communication with a central figure in every remote village, linked to the school, ensuring that online lessons were transmitted via offline mechanisms such as transcribing by hand and handing over the written transcripts to the children's families, in a kind of chain transmission that harked of the ancient postal system of the pre-modern era. Some teachers took the added initiative and taught the children one-on-one, traveling to their individual houses when needed. Many students have enthusiastically and actively participated in friendly competitions where they were asked to submit online entries (via video or photo sharing). The topics included handwriting in English, Nepali and Lepcha, arts and crafts, singing, dancing, speech and elocution, drawing and sketching.

This book arose in part from a short teacher's training course I and my husband offered the teachers of Mayel Lyang who were rather enchanted by the opportunity to learn in an open-ended dialectical manner, and collectively and individually discuss and witness the inter connectivity of a variety of sub-areas and topics from across disciplines. The course became an unforeseen opportunity to learn elementary mathematics in an integrated sense, be it contemplating patterns and number theory, the relevance of literature, art and architecture in mathematics, or examining the *sutras* of Vedic mathematics. The teachers themselves regained the joy of learning, buoyed by the prospect of new discoveries with a unique, inclusive and unrestricted take on education itself.

Ultimately, Mayel Lyang Academy and CMD is poising itself to meet the challenges of a highly uncertain future that may involve upheavals of a socio-economic and political nature. The students, teachers and administrators, guides and staff are all gathered together in addressing these uncommon circumstances. If nothing else, this experiment in setting up a school for rural tribal children has demonstrated that a sense of place and gathering in small groups, preferably with the privilege of access to plenty of land and unspoiled natural beauty, is imperative for cultivating a spirit of inquiry and for the flourishing of intelligent thought. In times where pure water, pure soil, and pure air are at a premium, it is all the more essential to include these natural God-given elements, metaphorically speaking, as important ingredients towards the search for community driven holistic education.

"Home is the first school for us all, a school with no fixed curriculum, no quality control, no examinations, no teacher training."

—Charles Handy

Four students from Brockwood Park School, together with a mature student and their teacher, emerged from the immigration area to a warm welcome by their hosts. Thus commenced the second innings of the Calcutta Academy, an appellation denoting both the venue and the root word *Akademikos*: a place where inquiring minds, led by Plato, assembled many aeons ago as *philosophers*, i.e. lovers of wisdom. Alas, the fairly capacious ten-seater Tata Sumo car proved to be a bit cramped after everybody's luggage was heaped onto the back row of seats, barely leaving four spaces in the middle row for the group of six. As one would imagine, the ride to the neighbourhood of Salt Lake in Calcutta, which the Brockwood guests would call home for the entire month, was nothing short of a roller coaster experience. It was indeed such a long journey, all for like-minded seekers to gather and understand, study and get to the essence of each subject that they chose to explore together. This exploration was purely for its own sake, without any diploma or paper acknowledgement of their efforts or, for that matter, any particular external incentive.

Before any of us could embark on our lofty endeavours, the visitors had to first overcome jet lag, make their acquaintance with mosquitoes, learn to sleep through the mad cacophony of yelping dogs that decided to play their symphony orchestra only past midnight, tolerate the endless *thud-thud* of an adjacent construction site, and ignore the ceaseless drumming practice for the annual Sports Day in the opposite park. After all, the students had arrived at a city *in extremis*, ever evolving, ever decadent, and yet all-encompassing.

The group brought with it its own energy, flavour, and particular chemistry, both between the individuals and as a collective. The expertise of the three teachers, and the mature student, who was attending his second Calcutta Academy, covered a wide range of subjects. Their competence spread from mathematics, physics and philosophy to literature, history, art and music *inter alia*. Our day seamlessly moved from sessions of mathematics and philosophy in the mornings—ranging from syllabus-related directed examination study to questioning the nature of trigonometry or of numbers themselves — to straying into history or art, and music theory in the latter half of the day and their link to mathematics. A simple home-cooked vegetarian lunch, in keeping with the strict vegetarian or vegan ethos of the school, served as a break.

The entire group was interested in learning together, with a serious effort at unearthing our individual likes, dislikes, prejudices and predilections towards various subjects or topics. Why had mathematics been so alienating in their past experiences? Was it the result of a bad classroom experience? What are the big fundamental questions that underlie all learning? Are subjects connected, or do we end up super specialising the topics until there is nothing left? Quite like the elusive pursuit for the perfectly sharpened pencil tip ... the pencil ultimately disappears.

We spent several weeks immersed in various subjects, exploring the core of each problem that we encountered - ruminating over ideas, theories, and philosophical bases of fundamental questions about nature and our purpose on this planet. In addition to examining specific topics, we also made an attempt to link seemingly disparate areas like mathematics and art and were pleasantly surprised to uncover the mathematical art of M. C. Escher whose paintings had inspired many a mathematician. While the journey into mathematics began with trigonometry and calculus, it broke free of convention with a tie-up class with art. We explored Escher's paintings and patterns and created our own, inspired by images and lectures. This gave an impetus to our understanding of mathematics, and the students became that much more eager to explore the scope of the subject beyond the classroom, even venturing on to atypical areas like *Vedic* mathematics.

The philosophy class was held in the nearby park, a short, pleasant walk from the house and a welcome change of scene —a perfect setting to allow the mind to soar up to the sky. The main idea circling in our minds was to dwell upon the nature of mathematics. An essay by Martin Heidegger, '*Modern Science, Metaphysics and Mathematics*,' formed the basis for our animated and fruitful discussions. The history class was wide in scope, but mainly covered European history. The demarcation between subjects became naturally blurred, in a way that seeks to clarify rather than befuddle. The music theory class dealt with the arrival of tempered tuning in modern music and the special way that alters the natural mathematical musical ratios.

Our typical day began with a yoga session on the rooftop among the twittering avian visitors from near and far: crows, canaries, budgerigars, orioles and sparrows, who left their commentary with us. A few students preferred a run around the park grounds opposite the house, sometimes indulging in a mock game of football with local amateur players. Before our lessons began, a couple of students would slip into the neighbourhood market, with its makeshift stalls skirting the entrance, to start the morning off with a drink from medium-sized green coconuts. Happy debating with the vegetable vendors lining the park would ensue, and produce from the nearby villages would be brought in soon after. A few students befriended a roadside tailor hawking his stitching and mending skills. When the academy was over, the students left with all their torn clothes darned, many of their oversized skirts or blouses altered to fit perfectly, and a promise to the tailor that they would somehow stay in touch.

The warmth arising out of thus living and learning together, eating south Indian *idlis* and *dosas* for breakfast, playing with our then eleven-month-old toddler, intensely engaging in bargaining for an insanely underpriced deal at the hand-loom emporium markets, sipping tea together at the Oxford Bookstore and remarking on what postcards to select and send home was a catalyst for learning without effort. Most evenings were laced with creative joy, and on any day one could spot the students and teachers alike scattered about the house, pencils and drawing books in hand, attentively painting their subject, or immersed in a book, or scribbling notes. From the sheer volume of drawings: sketches, portraits, patterned tiles (some akin to

Escher's etchings), water colours, charcoal drawings and ink paintings, one could surmise how everyone's creative instincts had been stimulated.

Many evenings ended with a movie session, and each film selected became, quite by accident, a visual continuation of the day's lessons. For instance, in *Chihwaseon (2002),* a Korean epic drama about a great painter of the late Josean era, one learned that madness and artistic genius were both essential for his craft. The film became an instant favourite with everyone, as it resonated with their own individual vision and artistic pursuits. Watercolour murals drawn during an extended pre-dinner session found their reflection in Hayao Miyazaki's hand-drawn watercolour animations. Bookish discussions on Mughal history from the afternoon's class were seen through the lens of Bollywood that same evening — in the biopic *Jodhaa Akbar (2008)*, inspired by the story of the great Mughal Emperor Akbar and his Rajput Hindu princess bride, Jodhabai. Books became our best friends, and the students appeared to quietly devour reams of writing — novels by Dostoevsky to an encyclopaedic treatise on how the brain works.

Attempts like the Calcutta Academy raise questions as to their apparent measure of *success*. Clearly, learning becomes effortless once the pressure to perform or be tested is lifted; more importantly, learning itself, for its own sake, provides the only incentive, one that suggests a deeper inquiry into the essence of every subject that we touch upon. One ingredient we found that was essential to learning without fear or effort, learning with delight, was again, *leisure*. The bustle of a school term, a long to-do list, and the general insanity of modern life rob us of this very basic quality. Did we then, through our simple living and unfettered thinking, discover what it takes to be fully able to learn: a state of leisure in togetherness, in the truest sense?

A wider question that undergirds this attempt to bring teachers and learners together in one place, without any sense of competition or external motive besides the impetus of learning itself, is: Why do we not experience this sense of togetherness in our typical classroom setting? Is learning and therefore any kind of exploration of mind related to friendship first? What about place - the place where we learn must be indelibly connected to how we learn. In our Cartesian culture of thinking we are not allowed to acknowledge the overarching importance of where we learn. Without an atmosphere of love, can the love of wisdom (*philosophia*) develop? Can we become excellent at a subject without the sense of competition that is often an automatic accompaniment to most academic environments? Are we then saying that comparison and a sense of mutual mistrust of fellow learners obstruct true understanding? These questions are open-ended and meant as an exploration of the possibilities of study and of the art of learning.

What the academy showed me personally, at many levels, was the magical effect of a natural camaraderie arising out of friendship and a common experience and exploration of place - in this case an overcrowded city. The small class size allowed for a deeper layer of bonding and interaction — ingredients that surely helped to cultivate a collective spirit of inquiry with complete abandon. No one was watching the clocks tick. The unrestrained nature of 'classes' meant that a topic could be unhurriedly obsessed upon for hours, allowing for many angles and aspects to be illuminated without a 'cutting off' at the end of a period that so often typifies ordinary classroom set ups.

The main motif for the effectiveness or success of our endeavours — if at all one can make such an assessment — were the drawings that emerged as a marriage between art, mathematics and philosophy, tying various themes into a collective whole. An image that comes to mind, itself a subject for the painter's palette, was of the students sitting around the dining table at night, each engrossed in their individual pursuits and yet threaded together in friendship. Ultimately, the right kind of education, and the re-structuring of mathematical learning as a whole would involve the proximity of nature, the blending of the arts and a seeking of friendship with a movement towards creative understanding. The title of this book reflects this very search and in my own meandering way I make an attempt to weave a story through this journey of mathematics towards an integrated comprehension of life itself.

At last I turn to my currently ongoing situation which, like the CMD stay, arose from unplanned circumstances — we ended up staying indefinitely at an organic farm and homestay run by dear friends after they extended a warm invitation to visit them during the 'off season.' As India's second more deadly wave of the covid-19 pandemic raged across the nation, the uncertainties that lay before us became starkly evident. Incidentally, our two children aged 6 and 11 have hardly had regular schooling, not so much on account of ideology but warranted by our extraneous circumstances which included taking care of my elderly and ailing parents.

The village of Phuldanga where the farm is situated borders the university town of Shantiniketan, home to Tagore's Viswa Bharati University and to so many Bengali intellectuals, mostly migrants from Calcutta — hoping to escape the city with the promise of a less frantic retirement. As unexpected luck would have it, everything grows in this farm-stay; fruit trees laden with ripening mangoes, lychees, papayas among a host of others — exotic species like jackfruit, starfruit, *batabi lebu* or pomelo (resembling grapefruit), *tepa cool* or Indian plum, and rose apple. Many of the other fruit trees I have never heard of before. They all had the dubious fate of dangling from the branches with happy abandon one day, and being turned into jam for the breakfast table the next! As though nature was suggesting that even in the bleakest of times, when death and uncertainty were knocking at the door everywhere, bounty and beauty were still around every corner if only one cared to look. The gentle healing quality, grace and largess of nature was still ever present. The real challenge was to keep my young children entertained and happy despite the stringent restrictions on movement. How do they continue to learn and spend some time away from their addictive mobile phones and computer screens?

What followed over the next few months was engagement and dialogue where learning was made central to our everyday existence, naturally arising out of daily tasks - simple acts like cooking, fruit picking, tending to the vegetable and herb garden and *witnessing*. Every day had reams of excitement stored up, even in the silences of the beating heat of the day. From observing how the parched red earth could metamorphose into satisfied soil after a few days of rain, to watching fractal designs being showed off by flashing lightning in the evenings, to conversations about cyclone patterns when the dreaded 'Yaas' cyclone struck mid-way (sparing this region, despite wreaking havoc in other parts of Bengal and the adjacent state of Orissa), it seemed *mathematics* wove its way into our lives rather deftly.

I discovered quite to my amazement that even the most cutting-edge ideas and concepts that involve considerable mathematical insight and advanced computational skills to get right are readily accessible, in spirit, to children who might not yet be able to read or write (such as mine!). Among the most difficult mathematical predictions to make is the weather. Modeling weather patterns purely in terms of the 'butterfly effect' has not proven to be as fruitful in the past. In chaos theory, the butterfly effect is the over-sensitivity of initial conditions on the evolution of non-linear systems — a small change in one state leads to a large change at a later time. They say, metaphorically, the effect of a butterfly flapping its wings could lead to the generation of a tornado, weeks later, in a different location. The accuracy of weather predictions has been greatly enhanced by including data — massive amounts of it — to shape the outcome of initial patterns. Explaining these concepts to small children, however keen they may be, can prove to be an uphill task — but only if you think so. Why not try? This is exactly what I did when the right context arose. Two million people had to be relocated before the tropical cyclone Yaas whirled its way into Bengal and Orissa. On the expected day of landfall, the children were nervously excited, eagerly noticing the agitated birds acting as the primary soothsayers of dangers to come.

At once, my son checked some satellite pictures and noticed that cyclones always appear as spirals when observed from above. He remembered noticing the same spiral effect in a bathtub sinkhole, where the vortex forms and spins rapidly as the water is released. And what is the link to galaxies that also have spiral arms? Both children learned about fluid flow and turbulence just by peeking from the balcony and glimpsing at the sombre skies. The swirling dark clouds providing their first lesson in the spiraling behaviour of wind and moisture — a stochastic dynamical system that nevertheless had much order in its chaos.

The eerie calm after the storm allowed us to venture out into the garden again. We noticed that we had literally received a 'mango windfall.' Hundreds of kilos of mangoes in various states of ripening meant that the next month or two would be replete with mango shakes, mango chutneys and jams, mango souffles, mango *achar* and mango fruit salads. I asked my children a question to ruminate over: why does shaking a tree make fruit fall? The trunk is in the middle, and the fruits are peripheral to it — so how does the force distribute itself...? Of course, I had to skirt the uncomfortable question of what 'force' is in the first place.

Another small exercise that involved taking the children on a simple garden tour and picking vegetables for the evening meal, opened their eyes to where the food they ate came from. Not only that, they were alerted to their mathematical properties as well. Observing the cross-sections of okra or bhindi as it is locally called, or the sturdy tomato, cut across the middle, revealed the same 'five' sectioned pattern. Again, what was going on? As my son uprooted onions, he cried out how he never knew it was an underground vegetable like the potato. My daughter squealed with delight when she noticed that the jasmine flowers she had plucked for her wicker basket, appeared in a heart shaped homemade soap later in the day and could be merrily consumed as herbal tea that very evening. The hibiscus flower with its white and red trumpet-shaped petals also obeyed the Fibonacci-Virahanka number system. The leaves had cleverly arranged themselves in such a way as to maximize the exposure to sunlight; twisting appropriately in degrees that were linked to the golden ratio. The hibiscus flower is frequently used as a beverage, known also for its high vitamin C content and healthful properties. Sukanya, our hostess, creatively uses these flowers in a variety of ways: for extracting oil, as tea, as a salad mix, in homemade soaps and shampoos and as a vegetable dye in batik printing for the clothes she designs. Inspired by the variety of the flora and fauna present in this relatively small acre and a half property, what was amazing was the seemingly infinite bounty that could arise from a farm of a modest size. Nature was not only eager to provide all the meals, the vegetables, fruits and herbs could be used for anything and everything — as medicine, as clothing dyes, in shampoos, creams and lotions, and even toothpaste. Both rational and irrational numbers were present in the arrangement of petals and leaves, in the cross sections of fruit and in the patterns of the weather. *In praxis* this bounteous garden of Eden has demonstrated the most subtle of mathematical lessons to be absorbed and assimilated without the typical classroom equipment that more often than not causes a sense of alienation.

Years ago, my undergraduate thesis, which has thenceforth spent its time in the bottom of a dusty drawer, involved the mathematical study of wrinkle patterns in elastic substrates; examining how forces distribute themselves in the interior when small changes are made in the periphery of an elastic substrate. The applications for such problems apparently ranged from the wrinkle distribution in airplane covers to the after effects of plastic surgery. My daughter became intrigued by the material properties of the natural fibres and materials she came upon in the little workshop and studio that happened to be Sukanya's workplace where she designed, stitched, wove and knitted clothing, ranging from children's dresses to batik printed saris. More than the creative labour that goes into the production of each clothing that is essentially a work of art, my daughter became interested in why some of the saris were painted with motifs and designs, butterflies, herons, swans, that progressively numbered 1, 2, 3, 5 and 8 as they were painted onto the *anchal* of the sari. The same recognisable numbers that show up repeatedly. Sukanya had no answer to the question as she previously had not noticed the pattern in her own designs and handiwork. Her pursuit was for beauty alone. So how do these arrangements come about when the artists themselves are unaware of such symmetries?

The exclusive cultivation of mathematical technique and syllabus oriented instruction in the teaching of mathematics might lead towards livelihood, but skirts the overall comprehension of the majesty of mathematical understanding — in that ancient sense of the *mathemata:* the already given gifts of life that we remember through the act of learning. The chapters covered in this book are an attempt to discover how mathematics is informed through the multifarious areas of learning and human endeavour, through observation, through co-operation, through building, dwelling and thinking together.

Going back to Galileo whose education was in the humanities and the arts, both of which collectively shaped his mathematics, we find that his intellectual inheritance was coloured by the holism of his widespread access to the genius of the late Middle ages and the early Renaissance. Much in the same way, we hope to re-attend to the broad philosophical and artistic foundations of the mathematical sciences, knitting together so many threads of thought, in order to re-imagine and rebirth a holistic and natural mathematics for a new era of learning.

AFTERWORD BY DR. VALENTIN GERLIER

The Renaissance philosopher, theologian and mathematician Nicholas of Cusa once taught that genuine insight occurs when opposite things are seen to mysteriously coincide, and that, consequently, the journey to wisdom is a path on which ever-greater knowledge is brought into accord with ever-greater ignorance. This wise, witty and splendid treasure-trove of a book performs just such an amazing journey: by showing us ever more clearly what a wondrous thing mathematics is, it leaves us utterly open to discovering what a mysterious thing mathematics might be. Dr. Sen's witty, perceptive, energetic interdisciplinary aptitude and her ability to frame her thoughts in a lucid, relaxed autobiographical style invite us on a re-oriented journey towards the heart of being, where learning maths coincides with imagining, where knowledge is gathered only as an attendant to friendship with and reverence to things and to beings, where playfulness is always serious and seriousness slippery and playful. Such qualities, we are shown, arise also in the patterns of the universe, of Being itself, and are not arbitrarily imposed by the human mind, as an aesthetic or scientific wish-fulfilment imposed upon some chaotic, indifferent stuff 'out there.' It seems as though the path-finding, serpentine journey performed in this book is itself a mathematical pattern of a sort, where the author does not merely go through an enumeration of items of information, but continuously delights and participates in her enigmatic, yet real, findings: equations are moments of friendship, fractions moments of relationship. The mathematical, then, is not the purview of the methodical: it continuously yet unpredictably shows itself through the imaginative, the interpretive, the creative aspects of a journey into reality.

Indeed, the book is a challenge to those who believe truth and 'factuality' should reflect the rigorous sobriety of arithmetic poise, or emanate the reassuring whiff of dry, laboratory-like procedure we all have been taught is what we do when we transform things into numbers. In fact, as Dr. Sen shows us, it is not possible to transform the world into arithmetic (*pace* Galileo) since numbers, it turns out, are not at our disposal: they too are things and as such have the dynamic, withdrawn elusiveness and presence that belongs to thinghood — rather than the alphabet of reason, they turn out to be the magic rings of truth. Here, mathematics beckons as a wondrous realm, like a Celtic Otherworld, whose longitude and latitude intersect with the multiple horizons and temporalities of natural and human creativity. To glimpse the elusive essence of maths, one must have a taste for the adventurous, the unpredictable and even the humorous — the very opposite to the indifferent application of rational method. Indeed, one must develop a feel for how mathematical truth 'shows up,' attending without impatiently seeking to 'grasp', when we traverse, encounter or dwell with the amazing, textured interwovenness of the natural world. We are also taught to gently attend to poetry, art, or music too, where patterns dance with infinity and where structure speeds the flight of freedom. Instead of the predictable, the constant, the controllable, we are asked to watch for the unforeseen, the irregular, that which does not submit to the cool gaze of reason but, fairy-like, glimmers at its edges. This is true of the patterns on shells we discover on empty ocean beaches, but it also true of the dynamic and spontaneous energies of authentic friendship.

As someone whose own journey is to seek to creatively articulate facets of this implicate order, I am left with some striking metaphysical considerations which flow from Dr. Sen's inventive and imaginative approach. It would be easy to simply conclude, as many geometrically-inclined philosophers over the centuries might seem to have done, that numbers are signs which perform a clear metaphysical job: they offer a vocabulary for the underlying, patterned structure of the real, some sort of sterile or frozen 'ideal forms,' which material reality only reflects in shadowy ways. That the realm of the intelligible should be cold and unfeeling, and the realm of the sensible mere shadows of its cutting logic, is a view wrongly attributed to many a philosophical schools, starting from Pythagoras and Plato. Herself a learned and astute philosopher, Dr. Sen's mathematical sensitivities point in a much more fruitful and nourishing direction – that the patterns of this world, embodied in dance, sounded in words of poets, woven by storytellers, painted by artists and discovered by genuine teachers along with their students, are ways in which *numbers too* become vitalised. Through such productions, numbers show up increased in their being: they become more numberlike in their number-ness. The sphere of mathematics, then, is not merely exemplified in but *nourished* by the arts – numbers become more real, gather more presence, partake in more thinghood the more they are imaginatively instantiated. It is possible then, that the stars in Vincent's painting, the patterns of the Shiraz Mosque or the wistful quatrains of Omar Khayyam are all also bestowing gifts to the numbers which underwrite their patterns: the relationship between implicate and explicate is therefore mutual. Poets, artists and philosophers increase the being of numbers, add to their reality, just as much as number make their creations real. They all engage in mathematical speculation and discovery.

It seems, then, that artists and mathematicians alike engage in the same practices: to show the un-showable, the hidden pattern of reality; or better put, to provide dwelling places in which such order can show itself in its own way. If number are real things, as this book constantly reminds us, and not merely techno-rational concepts, so too are the shapes and structures in which numbers dwell. Hence, a second metaphysical consideration is that *series* of numbers discussed in this book do not operate under the dominance of reassuring predictability — which we are so often taught mathematics bestows — but joins together the predictable with the unpredictable; a veritable coincidence of opposites. Of course, the number series guarantees the gift of repetition, just as much it is brought into being by it. Yet at the same time, the series offer surprising moments in which it suddenly becomes a kaleidoscopic entrance into reality, a totally unforeseen moment that reveals the open nature of the mathematical sphere. To return an example so perceptively discussed in the pages above, the Virahanka-Fibonacci series does not develop into a flat infinity of organised and sterile repetition. The numbers do not merely seem to 'stand there' in repetitive horizontality but gradually become spirited away, through folds and fissures in being, towards a greater actualisation of their essence — 'spiralling outward' as Dr. Sen puts it. This is *participation* in a series which also paradoxically gives a new kind of creative potency to the numbers 1, 2, 3, 5, 8, 13 etc. which compose it. As numbers participate in shape, so do shapes in turn shape the numbers which make up their unities. By bestowing to each member of the series this additional, circuitous, dynamic mode of being, numbers are released from flat, linear infinity into spiralling dynamic eternity. Hence, this serial way of encountering number is not merely a way of counting but of *accounting*, of telling the tale of Being: it is a way of thinking, of imagining, almost a mode of reverential *invoking* the sacred order of things. As such, it is also an invitation to re-imagine our dwelling the world: for the patterns of semblance discovered in pineapples, pine cones and daisies are not merely material for comparative interest: they indicate a real cosmic *sharing-in*. Such sharing suggests that quiet meadow flowers growing in the rainy Cambridgeshire countryside, which have inspired many a poem, quiet musing or a sudden holding of hands, are embraced in a spiral which also gathers in its sweep the wild fruit emerging by serpentine rivers moving through

Southern Brazil and Paraguay, chiming in with the Inca, Mayan and Aztec cultures that cultivated it. How do such different worlds partake in the same spirals? What might this teach us about our world and the way it is constituted? That is left for mathematicians and poets to articulate.

Hence, the series not only presents us with the interweaving of nature and culture, so popular among philosophers today, but more importantly opens windows and thresholds of kinship between apparently unrelated things. Like in the beautifully sonorous *cynghanedd* Welsh poetry, for example, where seemingly distant beings are brought into kinship through the play of assonances. Numbers touching one another in this way articulate the magical ligatures of the real: there is something of a voice there, a kind of speech at once harmonious and bewildering, at once a soothing charm and a perplexing riddle. We are invited to see the mysterious intertwining not merely between the daisy and the pineapple, but between the worlds in which these grow and have their being — again, mathematics is an Otherworld, a threshold to magical journeys where dimensions of reality appear mysteriously together, yet distinct. To 'think' the Viharanka-Fibonacci series may well mean to 'think' and 'see' the pine cone and the daisy, the pineapple and the lily as evoking the spectral family to which they belong, and enable to re-orient how we see, feel and understand what patterns and orders are asking us to participate in them: in other words, mathematics belongs to a family that includes poetics, aesthetics, philosophy and theology, but also ecology and ethics. To give but one example, the nature of the series that these plants and beings share in also teach us something about 'monocultures': in their *interruption* of the series, a row of industrially farmed pineapples or sunflowers grown for-cut flower use, distinctly show how the monologic and sterile repetitiousness of 'one' interrupts the hidden order mathematics seeks to actualise. In such practice, 'one' is not the Parmenidian, infinity-bestowing unit, but the endless repetition of a solitary self, bereft of its mathematical kin, its life-giving spiral. It would seem, from this point of view, that the obsession with identical repetition in agriculture also sheds light on the dangers of the profoundly unmathematical culture which underwrites it. In the light of the mathematics this book gives us, we are called to re-imagine our way of dwelling with the earth.

Hence, As Dr. Sen teaches us, patterns are unfolding aspects of a dynamic hidden order, and it is this sensitivity that she wishes to awaken in education. The education she wants to bring into being is a world away from over-populated classrooms, dry mark-schemes, exam preparation, aims and objectives, managerial jargon and the deadening hegemony of the fancy acronym. Such education, like the monocultures of fruit and flowers, is unmathematical, and has transformed numbers and patterns into tools for techno-rational control. Conversely, education 'when man does not interfere' is mathematical education, simply because it allows us to learn that which make itself manifest and comes upon us by itself. Yet these showings cannot happen in an uncommunicative wilderness, but must be offered a dwelling place: a sense of community gathered together in the name of the wondrous, inexplicable yet life-bestowing order she continuously seeks to show. The many insight-laden educational instances she recounts are made possible in underlying atmospheres of friendship, of warm generosity blended with a gentle, abiding rigour towards discernment. To create such conditions — and be created by them! — one must learn a 'sense' for the perfume of a shape, the feel of a series, the taste of a pattern that the Real Intelligence breathes out of its own infinite store of goodness. As educators, it is this sense that also attunes us to the right numbers of students in a class, to when to build a structure and when to loosen it, when to contemplate a mathematical problem and when to contemplate the ray of sunlight that illuminates the numbers on the page, or the maple leaf from which both the page and the numbers take their being. Education becomes a matter of mathematical attunement in which our imagination is always completely participating, both creatively and attentively. As a movement into learned ignorance, the educator coincides what *at first* seemed opposites yet, as this wonderful book teaches us, are not: number and creativity, the mathematical and the imaginal.

—Dr. Valentin Gerlier
St Edmund's College,
Cambridge

WHY I WROTE THIS BOOK

This book arose out of my own personal journey in search of mathematical truth, both as a practising mathematician and physicist initially and later as a teacher of school and college students. Three main incidents have kick-started and defined much of this seeking and have coloured my own attitudes towards learning and teaching. Luckily, the first occurred reasonably early in my path during one particular lesson in a class on creative writing. The assignment was to write two separate essays on the same topic. One essay was to be deliberately functional and accurate but lacklustre. The second essay on the same subject had to be the most remarkable and engaging writing that one could come up with. I soon realised that a supposedly dull topic, say for instance 'the refrigerator,' could appear bland and technical or, in contrast, could come alive on the page and inspire the reader, and even have a transformational effect on them. I then understood that the same principle could apply to the learning or teaching of mathematics. But why didn't the learning of the subject have that kind of riveting effect on most students? Why is mathematics the most hated subject in school among a majority of pupils? Why is mathematics taught mechanically and not as a part of living? These questions leading to a panoply of related questions began to flood my mind.

The second epiphany occurred during a lecture on the rings of Saturn delivered by Professor Carl Murray of Queen Mary's University, London, in 2006. He spoke about the images of Saturn's narrow and contorted F ring returned by the Cassini spacecraft indicating 'structure' and 'features' related to the perturbing effect of its natural satellite Prometheus. The perturbing effects of the inner shepherding satellite seemed to introduce channels through the F ring and a 'streamer,' a line of particles linking the satellite to the ring. Although I did not pick up many of the details of this talk, I do remember it having a transformative effect on me. I imagined how vibration could have a link to the arrangement of matter or could create material manifestation. So what was going on? The many rings of Saturn are separated from one another by distances that are somewhat dynamic. Do these distances of separation have any link to vibration, even to music? Was this the elusive proof for the music of the spheres? More important than the ratios of the spacing of Saturn's rings having structure and meaning, I realised that there was much work to be done in understanding the nature of mathematics and its link to matter; one could not separate vibration or the unseen from the revealed material manifestation or the seen. Was this structural spacing linked to the semi-tones of Indian classical music? I may have been day-dreaming, but it seemed to me that the structure of the interior rings of Saturn and their spacing had a connection to the ratios (semitones) found in ancient Indian classical music. Regardless of the details of this talk that left me somewhat emotionally paralysed, I understood how there was a dearth of astronomical, geological, natural and geophysical instruction in our curriculum. And including them in such a way as to form bridges between subjects – say between music and physics, mathematics and poetry or between the observations in nature and the classroom experience of arithmetic, would be highly illuminating. We needed a major re-haul in how we learned mathematics. We needed a more inclusive way, such that it encompassed everything. It was very important to connect the world that we see around us with the understanding of numbers, arithmetic, geometry and algebra. My later teaching experience confirmed what was required: a more holistic relationship with mathematics emerging from a keen observation of the natural world.

The third revelation was a matter of coincidence – or was it? My father's friend and colleague in the Indian Administrative Service happened to be married to the 'human computer' Shakuntala Devi and we all lived together in the same government quarters in 1970s Calcutta where I regularly interacted with her as a child. Later as a young adult I read some of her popular books and saw her interviews on television. I became intrigued by her astounding capacity and understanding of numbers and her ability to manipulate them in a way that defied belief. In 1980, she multiplied two random thirteen digit numbers, selected by the Department of Computing at Imperial College, London, and correctly gave their answer in 28 seconds – more or less the time taken to speak them out. This momentous event was recorded in the Guinness Book of World Records. More impressive than the capacity of individuals to achieve unheard of feats, was that this close and effortless relationship to numbers could be had at all. Was there a manner in which mathematics could be presented that would help us speak the innate language of mathematics? This would imply the development of a mathematical mother tongue that arises from being alert to the extraordinary beauty of structure and number that we see around us. The way mathematics is taught to us from the very outset hinders the dawning of a more natural, effortless relationship to the understanding of patterns and order in the universe. Could we, for a moment, experience the ease with which mathematics comes to a person like Shakuntala Devi? It is true that mathematics is not restricted to the mere world of numbers and encompasses form and geometrical thinking, not to mention its close marriage to the physics of material processes. Nonetheless, a natural facility with the building blocks of mathematics through numbers, and through the observation of patterns and natural phenomena expressing order – both implicate and explicate – will open us to the sheer beauty of mathematics in ways that are no longer restrictive or even tiresome, but rather liberating and exhilarating.

From the beginning I decided to teach in a manner that was interdisciplinary in style; hoping to make the subject approachable and relatable to everyday living. Even when teaching introductory topics to teenagers, like the concept of factorisation, I strove to provoke the students into opening up their minds to endless applications and possibilities of the simple act of multiplying any two numbers. Taking an easy example, 12 x 11 = 132. The product of 12 and 11 can be easily solved by hand but the reverse process of 132 giving us

12 and 11 as factors, is rather more difficult. This dichotomy of the ease of regular multiplication versus the difficulty of factorisation is brought home with a slightly more involved example of the prime numbers 101 and 103. Although it is straightforward to multiply 101 with 103 from the left to right of the equation, giving us 101 x 103 = 10403, the contrary process of finding the factors of 10403 is much harder. We exploit this paradox of the ease of multiplication or the difficulty of factorisation in the area of cryptography and banking. When I first presented this dichotomy to thirteen-year-olds it was not only an 'aha effect' for them, but awoke them into recognition. They already knew somewhere that factorising was more difficult, but it never struck them that it would have any applications in the 'real world.'

The main purpose of this book is not to merely find entertaining and natural ways to approach the subject of mathematics, but to also question whether mathematics is in fact a single subject. The different chapters in this book are devoted to disparate areas of study; disciplines that are, ordinarily, in themselves treated as wholly separate from mathematics – like art, poetry or literature, biological phenomena, and even friendship and community spirit. I have tried to lay emphasis on areas where they seem to merge with mathematics or at least act in tandem. Not only do we find a close relationship of subjects like art or literature with mathematics, we find that a lot of mathematics has sprouted from these areas. Islamic art for instance, had surely influenced, during his visits to the Alhambra in Spain, the mathematically profound artistic output of the modern artist M.C. Escher, who in turn influenced the well known contemporary mathematical physicist Roger Penrose in some of his mathematical works. From the earliest times of the Indus Valley Civilisation in the middle of the third millenium, art, religion, and geometry have had an intimate relationship. In trying to answer the perennial question of "Which came first, the chicken or the egg?" - Art or Mathematics - I hint at Art as the forebear. In other words, the chicken came first! Culture must predate the discovery or invention of mathematics.

While teaching an elementary mathematics class at Brockwood Park School, I divided the topics into sub areas, relating subjects as far removed from mathematics (in its current framework) as literature or poetry, art and architecture, music, cooking or dancing. Dance and movement involve angles and patterns, the learning of steps and motion related to rhythm and beats. It is no wonder then, that modern dance movements and choreography, such as the Laban dance choreography with its geometrical exploration of space, or even classical Bharatnatyam and kathak dance, have mathematics naturally embedded within them.

The chapters that appear in this book cover some of these topics, underlining the role of nature and the detection of patterns in varied contexts. Topics like Sacred Geometry and Mystical Numbers serve to whet our appetites for mathematics through its presence in unexpected places, for example in the context of spirituality. Art and poetry also emerge as fertile ground for the development of the subject of mathematics, but more in relation to an effortless 'seat' for mathematics – a template or foundation for the mathematical patterns to emerge from. There are areas that are equally at odds with the subject of mathematics like cooking or dancing that we do not attempt to cover in this book, primarily for lack of space in an attempt to keep this book readable and short.

Whenever I have used this interlinked approach to exploring mathematical ideas, particularly with young students, I have noticed their eyes light up in delightful recognition. Is it a re-cognising or re-learning of things that they already knew from within but never articulated? The seemingly haphazard evolution of mathematics must have also happened in this manner, particularly in earlier times when the demarcation between subjects and topics was not as clear-cut. One of the purposes of this work is to open up to the possibilities of mathematics, through its re-imagination, noting its penetration in nearly all aspects of our lives, even in something as far removed as *worship*. This book is a meditation on such possibilities of cross-fertilisation and also an opportunity to question our current educational system that leaves much wanting in terms of approachability and attractiveness. Why are we not drawn to mathematics, a subject of such immense reach? This and many such questions form the philosophical basis for this re-imagination. Since many topics in the book like art, poetry and nature's patterns have influenced the development of mathematics, can we devise a curriculum for ourselves that takes the subject as emergent from the Humanities, or from the art and act of living – from the acknowledgement of nature as our first classroom?

—***Ashna Sen,*** *Shantiniketan*

BIBLIOGRAPHY

1. Emily L. Doolittle, Bruno Gingras, Dominic M. Endres and W. Tecumseh Fitch, 'Overtone-based pitch selection in hermit thrush song: Unexpected convergence with scale construction in human music,' PNAS 18, 2014 111 (46) 16616-16621
2. David Bohm, 'Wholeness and the Implicate Order,' 1st edition, Routeledge and Kegan Paul, 1980
3. Galileo Galilei, 'Discourses and mathematical demonstrations relating to Two New Sciences,' published in 1638
4. A. M. Turing, "The Chemical Basis of Morphogenesis,' Philosophical Transactions of the Royal Society of London. Series B, Biological Sciences, Vol. 237, No. 641. (Aug. 14, 1952), pp. 37-72.
5. Benoit Mandelbrot, 'The Fractal Geometry of Nature,' W.H. Freeman and Co., 1982
6. Ashna Sen, 'High Reynolds number flow in a circular flexible pipe,' PhD thesis, University of Manchester, 2004
7. Thomas C. Hales, 'The Honeycomb Conjecture,' Discrete & Computational Geometry, 2001, 25:1-22
8. Alexander Marshack, 'The Mesolithic Bone: Ishango. The Roots of Civilization,' Ed. Alexander Marshack. New York: McGrew-Hill Book Company, 1972: 27–32.

9. Laurence Sigler, 'Fibonacci's Liber Abaci,' A Translation Into Modern English of Leonardo Pisano's Book of Calculation, 2012
10. Joseph Gwilt, 'The Architecture of Marcus Vitruvious Pollio in Ten Bocitoks,' translated from the Latin, Lockwood & Co., London, 1874
11. Takashi Ida, '"Vitruvian Man" by Leonardo Da Vinci and the Golden Ratio,' Advanced Ceramics Research Center, Nagoya Institute of Technology, Japan, 2012
12. Clara Ionescu, Alain Oustaloup, François Levron, Pierre Melchior, Jocelyn Sabatier, Robin De Keyser, 'A Model of the Lungs Based on Fractal Geometrical and Structural Properties,' Proceedings of the 15th IFAC Symposium on System Identification Saint-Malo, France, July 6-8, 2009
13. Manjul Bhargava, Fields Medal Symposium, Toronto, 2016: Patterns in Numbers and Nature, Princeton University
14. Mario Livio, 'The story of phi, the world's most astonishing number,' First trade paperback Ed., New York City: Broadway Books, (2002, 2003)
15. Arthur Benjamin, 'The Magic of Math: solving for x and figuring out why,' Basic Books, 2016
16. M.S. Sinith, Shikha Tripathi, K.V.V. Murthy, 'Raga recognition using Fibonacci series based pitch distribution in Indian Classical Music,' Article 107381 Journal of Applied Acoustics, 2020
17. Plato: Timaeus, Peter Kalkavage (translator), Focus Philosophical Library, Hackett Publishing Company, Inc., 2016
18. Keith Critchlow, 'Islamic Patterns: An analytical and cosmological approach,' Thames & Hudson, 1983
19. Iasef Md Rian, Jin-Ho Park, Hyung Uk Ahn, Dongkuk Chang, 'Fractal geometry as the synthesis of Hindu cosmology

in Kandariya Mahadev temple, Khajuraho,' Building and Environment, Volume 42, Issue 12, December 2007, Pages 4093-4107

20. Rethinasamy Kittappa, 'Kolam designs, a Dravidian art form, in Indus Valley Civilization,' www.researchgate.net: 10.13140/RG.2.2.14206.02885
21. Vijaya Nagarajan, 'Feeding a Thousand Souls: Women, Ritual and Ecology in India, an Exploration of the Kolam,' New York: Oxford University Press, 2019
22. Fukagawa Hidetoshi, Tony Rothman, 'Sacred Mathematics: Japanese Temple Geometry,' Hardcover, Princeton University Press, 2008
23. Cheng Dawei, 'Suanfa Tongzong,' published in China, 1592
24. Shigekiyo Muramatsu, 'Sanso,' published in 2011 in Japanese and held by 7 WorldCat member libraries worldwide
25. Kinjal Shah, 'What is the significance of the number 108 and 1008 in Jainism,' www.academia.edu
26. Bhante Henepola Gunaratana, 'Buddhist Suttas for Recitation: A companion for walking the Buddha's path,' Simon and Schuster, 2019
27. Arthur I. Miller, '137: Jung, Pauli and the Pursuit of a Scientific Obsession,' W.W. Norton & Co., 2010
28. Ian Stewart, https://www.britannica.com/topic/number-symbolism/Number-and-reality
29. Shah & Haleem (eds.), 'The Oxford Handbook of Qur'anic Studies,' Oxford University Press, 2020, pp581, 587-88,
30. Book of Genesis in the Old Testament in the Bible
31. Dolbeau, Romain. "The p196_mpi page," www.dolbeau.name
32. Deep Rooted Tree (Korean: 뿌리 깊은 나무; RR: Ppuri Gipeun Namu) is a 2011 South Korean television series

33. Caldwell, Chris. "Harvey Dubner". The Prime Pages. Retrieved 29 April 2013.

34. Joseph Milne, 'The Harmony of the Cosmos, the Soul, and Society in Plato,' Living in Harmony by The Harmony Project, 8th Aug 2019

35. Robert Kanigel, 'The Man Who Knew Infinity: A Life of the Genius Ramanujan,' Washington square press, 1992

36. The Man Who Knew Infinity (2015), film directed by Matt Brown

37. Philosophy, Between Poetry and Mathematics, talk by Alain Badiou, NYU center for the Humanities, https://www.youtube.com/watch?v=lMPtwWDxUB4, Date: October 26, 2017

38. https://archive.org/stream/AvicennasCanonOfMedicine/9670940-Canon-of-Medicine_djvu.txt (text translation of Avicenna's Canon of Medicine), From an early Persian MS. (Add. 27.261. f. 371b)

39. Leonid Zhigun and Abdul Bangura, https://www.researchgate.net/publication/336994930_Aleksandr_Sergeevich_Pushkin's_Eugene_Onegin_A_Fractal_Analysis, Aleksandr Sergeevich Pushkin's Eugene Onegin: A Fractal Analysis. 11. 56-83, (2018)

40. Journal of Humanistic Mathematics, Math in Seventeen Syllables, A Folder of Mathematical Haiku, Volume 8, Issue 1, 2018, selected by the editors

41. Scott Milross Buchanan, 'Poetry and Mathematics,' Lippincott, 1962

42. Scott Milross Buchanan, 'Symbolic Distance in Relation to Analogy and Fiction,' Trench, Trubner & Company Limited, 1932

43. Alice Major, 'Wordshapes and Rhymescapes: Translating translation symmetry into music and poetry,' Math Horizons, Vol 20., Feb 2013, pp 20-23, Taylor & Francis online

44. Amit Chaudhuri, Finding the Raga: An Improvisation on Indian Music, Penguin/Hamish Hamilton, 2021

45. Conversations with my Indian Classical music teacher Pandit Nabhodeep Chakraborty on the basics of Hindustani classical ragas and their structure

46. A. Seidenberg, 'The ritual origin of geometry,' Archive for the History of Exact Sciences. 1:488–527, 1975

47. F. Staal, 'Discovering the Vedas,' Penguin, New Delhi, 2008

48. S. Sinha, N. Yadav and M. Vahia, 'In Square Circle: Geometric Knowledge of the Indus Civilization,' Chapter 27, "Math Unlimited: Essays in Mathematics" (Eds. R. Sujatha, H. N. Ramaswamy and C. S. Yogananda), Science Publishers/CRC Press, 2012, pp 451-462

49. Issam El-Said and Ayse Parman, 'Geometric concepts in Islamic Art,' World of Islam Festival Publishing Company Limited, 1976

50. The optics of Ibn Al-Haytham, Books 1-111 On Direct Vision, translated with Introduction and Summary by A. I. Sabra, 1989, The Warburg Institute, London

51. Hans Belting, 'The Double Perspective: Arab Mathematics and Renaissance Art,' Pages 521-527, Published by Taylor &Francis online : 23 Aug 2010

52. J. V. Field, 'The Invention of Infinity: Mathematics and Art in the Renaissance,' Hardcover, Oxford University Press, 1997

53. https://www.wikiart.org/en/piero-della-francesca, Piero della Francesa's paintings in his book 'De Prospective Pingendi,' in the mid-1470s to 1480s, and possibly by about 1474

54. Alexandre Borovik, 'Shadows of the Truth: Metamathematics of Elementary Mathematics,' Publisher: American Mathematical Society, 2010

55. Martin Heidegger, 'Modern Science, Metaphysics and Mathematics,' Die Frage nach dem Ding (Tübingen: Max Niemeyer Verlag), pp. 50–83, 1962

All photos and figures that appear in the text are taken from the public domain. Chapter 11 is partly based on articles by me that appeared in the Journal of the Krishnamurti Schools and The Statesman Newspaper, Kolkata, (May 2011). A section of Chapter 12 is also based on my article from the 16th issue of the Journal of the Krishnamurti Schools.

Made in the USA
Monee, IL
04 November 2026

67386692R00115